THE GENE'S-E[YE VIEW] OF EVOLUTION

With the theory of evolution by natural selection, Charles Darwin showed how a purely mechanistic process can produce the design and diversity of the living world. In Darwin's original formulation, this was a theory about individual organisms. Over a century later, a subtle but radical shift in perspective emerged with the gene's-eye view of evolution in which natural selection was conceptualized as a struggle not between organisms but between genes for replication and transmission to the next generation. This viewpoint culminated with the publication of *The Selfish Gene* by Richard Dawkins (Oxford University Press, 1976) and is now commonly referred to as selfish gene thinking.

The gene's-eye view has subsequently played a central role in evolutionary biology, although it continues to attract controversy. The central aim of this accessible book is to show how the gene's-eye view differs from the traditional organismal account of evolution, trace its historical origins, clarify typical misunderstandings and, by using examples from contemporary experimental work, show why so many evolutionary biologists still consider it an indispensable heuristic. The book concludes by discussing how selfish gene thinking fits into ongoing debates in evolutionary biology, and what they tell us about the future of the gene's-eye view of evolution.

The Gene's-Eye View of Evolution is suitable for graduate-level students taking courses in evolutionary theory, behavioural ecology, and population genetics, as well as professional researchers in these fields. It will also appeal to a broader, interdisciplinary audience from the social sciences and humanities including philosophers and historians of science.

J. Arvid Ågren is a Research Associate at the Lerner Research Institute at the Cleveland Clinic and a Visiting Fellow at the Evolutionary Biology Centre at Uppsala University. His research focuses on genetic conflicts and he has published widely on their biology and implications for evolutionary theory. He studied evolutionary biology at the universities of Edinburgh and Toronto, and has held positions at Cornell and Harvard.

Praise for *The Gene's-Eye View of Evolution*:

'Arvid Ågren has undertaken the most meticulously thorough reading of the relevant literature that I have ever encountered. And he deploys an intelligent understanding to pull it into a coherent story. As if that wasn't enough, he gets it right.' -**Richard Dawkins**

'The idea of the selfish gene revolutionised evolutionary thinking and led to many new insights. But from the outset, it received strong criticism, not all of it baseless. In the first dedicated book on the topic, Arvid Ågren expertly sets out the power and nuances of the selfish gene concept. At times taking sides, at others leaving history to decide, he is always perceptive, scholarly, balanced, and good natured. Interwoven with asides on the principal players, this fine book succeeds in being both enlightening and engaging.' -**Andrew Bourke**, Professor of Evolutionary Biology, University of East Anglia, UK

'Since its inception in the 1970s, the "gene's eye view of evolution" has been a controversial idea in evolutionary biology. In this lucid and scholarly book, Arvid Ågren provides a masterful treatment of the intricate and often confusing debates over the value and limitations of the gene's eye view. I highly recommend his book to anyone seeking a deeper understanding of this important issue.' -**Samir Okasha**, Professor of Philosophy of Science, University of Bristol, UK

'This book's conversational style, clear presentation and well-planted surprises make it ideal for both general readers and students in a broad range of fields. The selfish gene is alive and well and continues to inspire and irritate, which is why we see gene level arguments of fans and critics alike in past and present debates. Best of all, as we follow the gene's eye view around in Agren's book, we find ourselves educated about current views in exciting subfields-from evolutionary systems theory to Major Transitions and Selfish Genetic Elements- and rewarded with a treasure trove of references.' -**Ullica Segerstrale**, Professor of Sociology, Illinois Institute of Technology, Chicago, USA

'Science needs ingenious points-of-view that help us understand the world. Few perspectives are more famous—or notorious—than that of the selfish gene. Merging biology and history of science, Ågren unravels its origins, explains why it is useful, and when its utility has been overstretched. Whether you're a fan or a critic, this is an essential guide to the gene's eye view.' -**Tobias Uller**, Professor of Evolutionary Biology, Department of Biology, Lund University, Sweden

THE GENE'S-EYE VIEW OF EVOLUTION

BY

J. ARVID ÅGREN

OXFORD
UNIVERSITY PRESS

OXFORD
UNIVERSITY PRESS

Great Clarendon Street, Oxford, OX2 6DP,
United Kingdom

Oxford University Press is a department of the University of Oxford.
It furthers the University's objective of excellence in research, scholarship,
and education by publishing worldwide. Oxford is a registered trade mark of
Oxford University Press in the UK and in certain other countries

First published 2021
First published in paperback 2023 (with corrections)

Impression: 1

Published in the United States of America by Oxford University Press
198 Madison Avenue, New York, NY 10016, United States of America

British Library Cataloguing in Publication Data
Data available

Library of Congress Cataloging in Publication Data
Data available

ISBN 978–0–19–886226–0 (Hbk.)
ISBN 978–0–19–287259–3 (Pbk.)

DOI: 10.1093/oso/9780198862260.001.0001

Printed and bound by
CPI Group (UK) Ltd, Croydon, CR0 4YY

For
my father, the biologist
and
my mother, the cultural historian

Preface

One of my biggest embarrassments in life is that I am such a poor naturalist. My botanical skills are distinctly average and my ornithological knowledge is downright appalling. Rather than a love of natural history, what attracted me to biology was a fascination with the logic of the theory of evolution by natural selection. No other theory explains so much with so little. It is truly deserving of the title of 'the single best idea anyone has ever had', as Daniel Dennett once put it. And in contrast with other great theories of science, like general relativity or quantum mechanics, it can be (mis-)understood by anyone. I have always been drawn to the conceptual issues of evolutionary biology, questions that my hard-nosed empirical colleagues would dismiss as too theoretical, too abstract, and, if they wanted to be really mean, too philosophical.

This is a book about one of those issues, the gene's-eye view of evolution. The book came about thanks to Francis Crick's Gossip Test. According to Crick, your true interests are revealed by what you gossip about. For me, that has long been the gene's-eye view, and the vituperative debate that has surrounded selfish genes for the past half-century. As this book will make clear, the story of the gene's-eye view deals with many abstract questions, but it also has innumerable empirical implications. It strikes right at the heart of the question of what evolution is, and how we go about studying it.

I have been thinking about the disagreements over the gene's-eye view for the past decade, ever since I moved to Toronto, Canada, to begin my graduate research. Arriving in Toronto after growing up in Sweden and receiving my undergraduate training in Scotland meant that I for the first time came in close contact with students from North America. One thing that struck me about my new colleagues was that they often had a very different perspective on theoretical issues in evolutionary biology than I did. To exaggerate and overgeneralize a bit: if when I was a teenager and expressed an interest in the big questions of evolutionary biology I was handed a book authored by Richard Dawkins, they had been given one of Stephen Jay Gould's. I learned a tremendous amount discussing these issues in the lecture halls, seminar rooms, and, especially, in the Graduate Student Union pub located right next to the Department of Ecology and Evolutionary Biology of the University of Toronto. The genesis of this book owes a lot to those intellectual sparring sessions.

More concretely, several people were instrumental in making this book a reality. In particular, David Haig, and his Fundamental Interconnectedness of All Things discussion group, provided an environment that encouraged tackling the foundational questions of our field. Few have thought more about the gene's-eye view than David and he has been reliable source of advice and support throughout this project.

I am grateful to a large number of colleagues who took the time to read and provide comments on my writing. For help with individual chapters, I thank Alan Grafen, Alister McGrath, Andrew Bourke, Anthony Edwards, Cédric Paternotte, Charles Goodnight, Dan Dennett, Dan Hartl, Denis Noble, Ellen Clarke, Erik Svensson, Jack Werren, Jim Mallett, John Durant, Jonathan Birch, Kevin Foster, Lutz Fromhage, Megan Frederickson, Michael Bentley, Michael Rodgers, Stephen Wright, Stu West, and Tim Lewens. Others, including David Barash, Brian and Deborah Charlesworth, Andy Clark, James Marshall, and Martin Nowak, answered questions and pointed me to resources that have been tremendously helpful. For reading the entire manuscript, at one point or another, I am indebted to Chinmay Sonawane, Cody McCoy, David Haig, Jon Ågren, Manus Patten, Richard Dawkins, Samir Okasha, Steve Stearns, and Tobias Uller.

Their comments greatly improved the text, clarified my thinking, and offered much needed encouragement. On occasion, they also saved me from some embarrassing misunderstandings. They did not always agree with my argument; sometimes they were just kind enough to explain why I was wrong. Any remaining mistakes are, of course, mine.

Several people helped me understand how the gene's-eye view has been received around the world. I am grateful for the insight from Adrian Stencel, Amitabh Joshi, Ehud Lamm, Ian Caldas, Israt Jahan, Jae Choe, Jun Otsuka, Kazuki Tsuji, Leonardo Campagna, Philippe Huneman, Snait Gissis, Sakura Osamu, and Victor Luque. Only a small portion of this topic made it into the book in the end, but I hope to return to it in the future.

It has been said that being published by Oxford University Press is like 'being married to a duchess: the honour is greater than the pleasure'. That has not been my experience. Ian Sherman and Charles Bath have been extremely supportive and helpful since the first day of this project. I am grateful to Spencer Barrett for introducing me to Ian, and to Spencer and Locke Rowe for helping me draft the proposal.

Two chapters are partly based on material from my previously published papers and I thank the publishers for letting me reuse some of the text. Chapter 4 draws on Ågren JA. 2018. The Hamiltonian view of social evolution. *Studies in History and Philosophy of Biological and Biomedical Sciences.* 68–69: 88–93. Chapter 5 incorporates material from Ågren JA. 2016. Selfish genetic elements and the gene's-eye view of evolution. *Current Zoology.* 62: 659–665 and Ågren JA and AG Clark. 2018. Selfish genetic elements. *PLoS Genetics.* 14: e1007700.

Throughout the time of writing I was financially supported by the Wenner-Gren Foundations, whose generous support I am very thankful for. I also indebted to the staff at the Ernst Mayr Library at the Museum of Comparative Zoology at Harvard for their help and assistance, especially for their ingenuity during the extraordinary circumstances of a global pandemic.

Finally, I am forever grateful to my wife, Utako, for her never-ending love and support. And for insisting that I write this book.

J. Arvid Ågren

March 2021

Contents

Introduction: A New Way to Read Nature

There really is something special about biology. The French bio-chemist and Nobel Prize winner Jacques Monod described its position among the sciences as simultaneously marginal and central (Monod 1970, p. xi). It is marginal, because its object of study—living organisms—are but a special case of chemistry and physics, contributing to only a minuscule part of the universe. Biology will never be the source of natural laws in the way physics is. At the same time, if, as Monod believed, the whole point of science is to understand humanity's place in the world, then biology is the most central of them all. No other field of study deals so directly with the question of who we are and how we got here in the first place.

The location of biology among the sciences means that its theories can never be just theories. They will always touch us in a deeper way than those in other subjects. This is especially true for the theory of evolution. On one level, it is simply a theory of how different rates in sex and death lead to different configurations of carbon molecules. On another, it is our story of creation. How we think about the theory of evolution therefore matters more than the way we think about other scientific theories.

The gene's-eye view is one way to think about evolution. In its original formulation, Charles Darwin's theory of evolution by natural selection was a theory about individual organisms. Individuals vary in

The Gene's-Eye View of Evolution. J. Arvid Ågren, Oxford University Press. © J. Arvid Ågren 2021.
DOI: 10.1093/oso/9780198862260.003.0001

heritable traits, and if any of them make the individual more likely to survive and reproduce, these traits will become more common in the population as the generations go by. The gene's-eye view represents a subtle but radical shift in perspective. Building on the insight from population genetics that evolutionary change can be described as the increase or decrease of certain genetic variants, it argues that evolution is best thought of from the perspective of genes. By this reasoning, organisms are nothing but temporary occurrences—present in one generation, gone in the next. And, as a consequence, organisms cannot be the ultimate beneficiary in evolutionary explanations.

Instead, this role is filled by the gene. Genes are considered immortal and they pass on their intact structure from generation to generation. This way of thinking is also called selfish gene thinking, because natural selection is conceptualized as a struggle between genes, usually through the effects they have on organisms, for replication and transmission to the next generation. Here, 'genes' is used in a somewhat lax way. The evolutionary struggle is not between different genes within the same organism (though, as will become clear, the gene's-eye view offers a powerful way to think about such genetic conflicts) but between different alleles of the same gene within a population.

The origin of the gene's-eye view involved many people, but two stand above the rest: the American George C. Williams (1926–2010) and the Brit Richard Dawkins (1941–). The idea was first explicitly laid out by Williams in *Adaptation and Natural Selection* (Williams 1966), and then 10 years later more forcefully by Dawkins in *The Selfish Gene* (Dawkins 1976). Few phrases in science have caught the imagination of laypeople and professionals alike the way that 'selfish gene' has done, and it changed how both groups thought about evolution and natural selection. Among both groups, the gene's-eye view has subsequently amassed both strong supporters and fierce critics.

The debate over the value of the gene's-eye view has raged for over half a century. It has pitted 20th century Darwinian aristocrats such as John Maynard Smith and W.D. Hamilton against Richard Lewontin and Stephen Jay Gould in the pages of *Nature* as well as those of *The*

New York Review of Books. Even today, commentators cannot agree. For example, in 2015 when the science historian Nathaniel Comfort reviewed the second volume of Dawkins's autobiography for *Nature* he referred to the gene's-eye view as 'looking increasingly like a twentieth-century construct' (Comfort 2015). In contrast, writing for the same journal only a few months later, the biologist turned journalist and businessman Matt Ridley concluded that 'no other explanation [for evolution] makes sense' (Ridley 2016). Professional biologists remain equally divided. Simon Conway Morris dismissed it as 'an exploded concept that was almost past its sell-by date as soon as it was popularized' (Conway Morris 2008, p. ix). Similarly, a senior colleague of mine who I showed an early draft of this manuscript wondered if he really would be a suitable person to provide feedback as he disagreed with 'virtually every aspect of the field' and as a result he has 'trouble separating their bad science from good faith attempts to describe it'. Yet, Andy Gardner called *The Selfish Gene* 'unequivocally the most important popular book on evolutionary biology of the 20th century' (Gardner 2016), a view shared by the Royal Society who in July 2017 announced that, after a public poll, *The Selfish Gene* had been voted 'the most inspiring science book of all time'.

Much of the discomfort over the gene's-eye view comes from its surrounding vocabulary. Genes are 'selfish', organisms mere 'survival machines', and bodies nothing but 'lumbering robots'. The philosopher Roger Scruton complained that these ideas made 'cynicism respectable and degeneracy chic' (Scruton 2017, p. 49). Along these lines, a commonly told story (meaning it is seemingly impossible to track down the original source) is that *The Selfish Gene* was the favourite book of the disgraced Enron CEO Jeffrey Skilling, and that he used the book to justify the exploitative cutthroat culture of the company.

But the debate over the gene's-eye view is about so much more. It overthrows our conception of familiar biological terms like gene, fitness, and organism. It brings to the forefront evolutionary biologists' peculiar habit of speaking of biological entities as having intentions, deploying strategies, and pursuing goals. It drills to the core of what we mean by causality in evolutionary explanations. The gene's-eye view has featured in many major debates in evolutionary biology over the past half century, including in technical disagreements over epistasis, heterozygote advantage, and inclusive fitness, and in public discussions over what it means to be human. This book is about all of that.

How to think like a selfish gene

The gene's-eye view occupies a peculiar position within theoretical biology. In his review of *The Selfish Gene*, W.D. Hamilton called it a 'new way to read nature' (Hamilton 1977). What does that mean? The gene's-eye view is not a concrete empirical hypothesis (though it certainly helps us come up with such) and it is not an all-encompassing mathematical framework (though general models can be constructed). Instead, it is a way to make sense of the biological world. Dawkins once described it as: 'a particular way of looking at animals and plants, and a particular way of wondering why they do the things they do' (Dawkins 1982a, p. 1). Put like that, one can easily see why it would be difficult to come up with experiments that would reject it. This could be viewed as a strength or as a weakness. Take for example, one of the most quoted passages from *The Selfish Gene*:

> Now they [genes] swarm in huge colonies, safe inside gigantic lumbering robots, sealed off from the outside world, communicating with it by tortuous indirect routes, manipulating it by remote control. They are in you and in me; they created us, body and mind; and their preservation is the ultimate rationale for our existence.   (Dawkins 1976, p. 25)

Dawkins's statement contains the essence of the gene's-eye view: while Earth is inhabited by organisms, what ultimately matters for evolution is the propagation of genes, a process that genes play an active part in. The gene's-eye view prompts us to look at biological phenomena and ask *cui bono?* Who benefits? (Dennett 1995, p. 325). The perspective treats adaptations as the central problem of evolutionary biology and argues that it is only by understanding that genes, not organisms or groups, are the ultimate beneficiaries of natural selection that they can be understood.

As Denis Noble has pointed out, however, this notion is unfalsifiable. The only empirical content of the above statement is that genes are located inside organisms (Noble 2006, pp. 12–13). Noble rewrites the statement, removing what he considers to be unsubstantiated speculation:

> Now they [genes] are trapped in huge colonies, locked inside highly intelligent beings, moulded by the outside world, communicating with it by complex processes, through which, blindly, as if by magic, function emerges. They are in you and me; we are the system that allows their code to be read; and their preservation is totally dependent on the joy we experience in reproducing ourselves. We are the ultimate rationale for their existence.   (Noble 2006, pp. 12–13).

In Noble's version the organism is in control; genes are passive prisoners. There is no clear cut empirical data or ingenious mathematical model that can distinguish Noble's version from Dawkins's.

To make sense of disagreements like this, Ullica Segerstrale has suggested that the right way to approach the gene's-eye view is with a logical rather than a literal mind (Segerstrale 2000, p. 261). By a logical mind, she meant that the gene's-eye view grew out of an intellectual environment where it was common to design, often quite elaborate, thought experiments to explore how evolution works. In the preface of *The Selfish Gene*, Dawkins makes this point explicitly when he describes the book as being 'designed to appeal to the imagination' (Dawkins 1976, p. xi). Segerstrale contrasts such a logical environment with the literal intellectual milieu that tends to characterize experimental and molecular biologists. Here, the focus is less on exploring possible theoretical scenarios and more on carefully describing the features of the biological system at hand. Someone trained in the logical tradition will have no problem asking themselves 'if I was a gene, what would I do in this situation?', whereas a biologist from a literal background will find the question absurd.

The emphasis on the logic of evolution and natural selection shares many features with how philosophers do their work. Segerstrale argues—as do Sterelny (2001, pp. 4–6), Ruse (2003, p. 164), and Noble (2006, p. 11)—that this philosophical approach to evolution is more common among British biologists than among their colleagues from other countries. This may at first seem strange, as it has been said of the British that they do not do philosophy. For example, Ernst Mayr complained about 'illiterate Anglo-Saxons' who had not read enough Hegel (Kohn 2004, p. 329), and Iris Murdoch described the picture of the world painted by English philosophers as one 'in which people play cricket, cook cakes, make simple decisions, remember their child-hood and go to the circus' (Murdoch 1953, p. 35). The debate over the gene's-eye view shows why Mayr and Murdoch were being unfair. Disagreements over the value of the ideas in *Adaptation and Natural Selection* and *The Selfish Gene* raised the bar for conceptual discussions of biological phenomena considerably and were instrumental in the emergence of the philosophy of biology as a distinct field of study.

This book is about how to think like a selfish gene. Darwin once described the term 'natural selection' as 'in some respects a bad one, as it seems to imply conscious choice; but this will be disregarded *after a little familiarity*' (Darwin 1868, p. 6; my emphasis). My goal is to provide a little familiarity to the gene's-eye view: what it is, where it came from, how it changed, and why it still evokes such strong emotions. In so doing, I will argue that the gene's-eye view is one of the most powerful heuristics or thinking tools there is in biology. Like all tools, however, to get the most out of it you must understand what task it what designed to solve.

Aims and outline of the book

I have written the book I wish already existed. There was no book-length treatment of the history and current state of the gene's-eye view. Instead, anyone looking for a systematic summary of the topic has typically had to scramble together papers and chapters from multiple sources. Furthermore, much of the debate about the value of the gene's-eye view and its role in modern evolutionary theory has been conducted in the philosophical literature, which has meant that many biologists are unaware of the many nuances of the case for and against selfish genes. Though things are improving, especially in evolutionary biology, it would be going too far to say that the attitude often attributed to Richard Feynman that 'philosophy of science is as useful to scientists as ornithology is to birds' has completely disappeared.

This books aims to teach the controversy. The term 'teach the controversy' has a bad reputation among evolutionary biologists, and for good reason. For the past decades it has been the cornerstone of the strategy applied by the Discovery Institute to discredit the status of evolutionary theory and to promote the teaching of intelligent design in American public schools. It played a key role in the Kitzmiller vs Dover Area School District trial of 2005, which ruled that intelligent design 'cannot uncouple itself from its creationist, and thus religious, antecedents', meaning that its teaching violated the constitutionally mandated separation of church and state. It is therefore ironic that the term was originally coined by a self-described secular liberal professor of English, Gerald Graff, with the aim of showing students that academic knowledge is the product of a dynamic process with disagreements that are settled through empirical observations and deductive reasoning (see Graff 1992). Personally, I have always enjoyed popular accounts of scholarly debates, whether about string theory in *The Trouble with Physics* (Smolin 2006) and *Not Even Wrong* (Woit 2006), or the role of mathematical modelling in economics and physics in *Economic Rules* (Rodrik 2015) and *Lost in Math* (Hossenfelder 2018), and I carefully followed the debate about participant–observer ethnography in the aftermath of the publication of *On the Run* (Goffman 2014). Learning the controversy is a thrilling way to be introduced to a new field and this book is written in that original Graffian spirit.

The book spans five core chapters. In Chapter 1, I outline the historical origins of the gene's-eye view. I argue that its intellectual nucleus stands on three legs. The first is the tradition that takes adaptation, the appearance of design in the living world, as the central problem that the theory of evolution must answer. This view of the field has been particularly prominent in the United Kingdom and I trace this back to the historically strong standing of natural theology in the country. The second leg is the emergence of population genetics during the development of the Modern Synthesis. J.B.S. Haldane, Sewall Wright, and R.A. Fisher all showed how evolution can be mathematically captured as a change in allele frequencies, which moved the attention of biologists from organisms to genes. I will argue that Fisher was especially important and that he introduced a novel conception of the environment fundamental to the gene's-eye view. The final leg is the levels of selection debate. This is the debate over whether natural selection can act on levels above or below the individual organism. In particular, both Williams and Dawkins identified the widespread, and in their view misguided, popularity of group selection as providing the direct impetus to write their books.

Both *Adaptation and Natural Selection* and *The Selfish Gene* received a lot of attention upon publication and the ensuing debate quickly resulted in a rather confusing vocabulary. In Chapter 2, I describe how the gene's-eye view uses familiar terms in unfamiliar ways. Most important of these terms is 'gene' itself, which is used in a somewhat abstract way, agnostic about any molecular details. I also outline how the gene's-eye view matured in light of the criticism it received and how the abstract use of the term gene was integral to one of the gene's-eye view's most fundamental claims: that evolution requires two entities, replicators and vehicles. I discuss how this formulation of selection and evolution fares relative to other general accounts, such as Lewontin's principles, and how this relationship has changed due to the growing interests in the major transitions. Finally, I evaluate the concept of memetics, the application of the gene's-eye view to cultural evolution and outline why it has failed to have the same influence there as it had on the study of organic evolution.

In Chapter 3, I tackle some of the most common criticisms of the gene's-eye view. I first discuss the use of intentional language and the anthropomorphizing involved in calling genes selfish. I show how this habit has greatly annoyed critics within biology and beyond, and that this debate is the most recent instantiation of the old conflict over whether teleological explanations have a place in biology. Next, I outline the criticisms that came from mathematical population geneticists. I revisit the classic cases of epistasis and heterozygous advantage to evaluate how the gene's-eye view handles interactions between genes, and whether it is bound to commit the so-called averaging fallacy. I also discuss the charge from developmental biologists that the gene's-eye view affords unwarranted amounts of causal power to genes in the process of development and in so doing ignores too many other interesting biological processes. Lastly in this chapter I deal with how the gene's-eye view fits into evolutionary biology's troubled relationship with the concepts of human nature and morality. This is a debate that raged intensely in the early days of the gene's-eye view but which has by now largely calmed down. Nevertheless, I show how a close reading of the views of Williams and Dawkins on these issues reveals interesting parallels with those of T.H. Huxley.

Chapter 4 examines the long and intimate association between the gene's-eye view and the work of W.D. Hamilton. Hamilton was crucial to the emergence of the gene's-eye view, especially through his work on inclusive fitness and what is now known as Hamilton's Rule. Hamilton's key insight was that individual organisms can affect the transmission of their genes not only through personal reproductive success but also through the success of close relatives. Inclusive fitness provides a way to view this process from the perspective of the individual organism, but it can also be seen from a gene's-eye view. Dawkins and others have repeatedly emphasized the formal equivalence of the two perspectives. Yet, as I will argue in this chapter, there is an under-appreciated potential tension between the two. I demonstrate how this tension shows up in both the disagreements over the value of inclusive fitness theory stemming from Nowak et al. (2010) and in Alan Grafen's ongoing Formal Darwinism Project. I end the chapter by discussing two recent attempts to resolve this tension.

The gene's-eye view initially earned its stripes by helping the field make crucial strides in the study of old problems in evolutionary biology, in particular related to social behaviour. It also helped launch new avenues of empirical research. Chapter 5 is dedicated to three such new areas. The first is extended phenotypes, which are examples of phenotypic effects that occur outside of the body in which the gene is located. The second area is greenbeard genes, which gets its name from the thought-experiment devised to show that for altruism to evolve it is the relatedness between the actor and the recipient at the locus underlying the altruistic behaviour that matters, not the genome-wide relatedness. Finally, selfish genetic elements are genetic elements that have the ability to promote their own transmission even if it comes at the expense of the fitness of the individual organism. I discuss the state of the current understanding of these phenomena and the role of the gene's-eye view in uncovering them. I also show how all three examples add ammunition to the gene's-eye view's attempt to undermine the centrality of the individual organism in evolutionary explanations by demonstrating how individual organisms may fail to maximize their own fitness.

The book is primarily aimed at graduate students in biology, upper year undergraduates, and others seeking an introduction to the gene's-eye view of evolution. I also hope that it will be useful to more senior people in evolutionary biology and neighbouring fields looking for a fresh perspective on a familiar debate, as well as to philosophers and historians of biology who are interested in how biologists view this part of their recent history. Any primer struggles to find the balance between being cohesive and being comprehensive. In writing this book, I have had two aims: firstly to produce a book that would be cohesive enough to offer a one-stop introduction to the many debates surrounding the gene's-eye view, and secondly to make it comprehensive enough to show just how vast and nuanced these debates have become and to provide a road map to anyone who wants to go on and further explore this rich literature on their own.

I

Historical Origins

1.1 Introduction

Several books and papers were crucial for the emergence of the gene's-eye view, but it culminated with the arrival of *The Selfish Gene* in the fall of 1976. The exact timing of the book's publication owes much to two defining conflicts of modern British history. The first was the industrial conflict between the UK National Union of Mineworkers and Edward Heath's Conservative government. In the 1970s, coal burning was responsible for the majority of British electricity production, and strike action by coal miners had led to serious power shortages. The situation climaxed on New Year's Day 1974 when the government introduced the so-called 'three-day week'. One consequence of the lack of power was that Richard Dawkins, then a young lecturer who had recently returned to Oxford after a stint as Assistant Professor at the University of California at Berkeley, stopped his research on cricket behaviour and turned to writing. When the conflict was over, and power returned to normal, one chapter was complete. It lay forgotten in a drawer until a sabbatical in 1975, which Dawkins spent at home in Oxford writing the next chapters of the book in a 'frenzy of creative energy' (Dawkins 2016, p. xiii).

The other was the Battle of Trafalgar. Michael Rodgers, the Oxford University Press editor who would eventually acquire the book, was first told about the project by the physicist Roger Elliott, one of Dawkins's colleagues at New College, Oxford, and then later came

The Gene's-Eye View of Evolution. J. Arvid Ågren, Oxford University Press. © J. Arvid Ågren 2021.
DOI: 10.1093/oso/9780198862260.003.0002

across it in an internal memo that reported a conversation between the head of the Academic Division at the Press, Dan Davin, and the philosopher Anthony Quinton who had mentioned that 'a New College man was writing a rather lively book on the function of the gene' (Rodgers 2017). Having read the first few chapters, Rodgers is said to have rung Dawkins, booming down the telephone 'I must have that book!' (Dawkins 2013a, p. 276). *The Selfish Gene* was published on 28 October 1976, having been delayed one week to avoid clashing with the release of *The Oxford Companion to Ships and the Sea*. In the UK, 21 October is Trafalgar Day, which celebrates the Royal Navy's victory over Spain and France at the Battle of Trafalgar in 1805 and, on recommendation of the W.H. Smith booksellers, naval books were given priority (M. Rodgers, personal communication).

Dawkins has repeatedly emphasized the debt that the views presented in *The Selfish Gene* owed to other biologists, in particular R.A. Fisher, John Maynard Smith, Robert Trivers, W.D. Hamilton, and George Williams. He asked Trivers to write the foreword to the book, described Fisher and Hamilton as the leading candidates for the title 'most distinguished Darwinian since Darwin' (Dawkins 2000) and said that the most important thing for a successful conference was that Maynard Smith would attend and that the venue have 'a large and convivial bar' (Dawkins 1993a). After *The Selfish Gene*, Williams's *Adaptation and Natural Selection* is the most important book in the gene's-eye view's origins story. In his foreword to the 2019 edition of that book, originally published 10 years before *The Selfish Gene*, Dawkins describes it as the most important book in evolutionary biology of the second half of the 20th century, on par with the 'canon' of the 1930s and 1940s, and that it was the only book that he would insist that all his students read during his time as an Oxford tutor (Dawkins 2019).

This chapter traces the history of the gene's-eye, arguing that three areas provided the intellectual foundation: adaptationism, population genetics, and the levels of selection debate. Adaptationism is the tradition that takes the appearance of design in the living world as the cardinal problem that a theory of evolution must be able to explain.

This tradition has been especially important in British biology and its through that intellectual milieu that the gene's-eye view emerged. The gene's-eye view treats adaptation as a special kind of problem that deserves a special kind of explanation. Key advocates of the gene's-eye view have often invoked the writings of the theologian William Paley, the author of *Natural Theology* (Paley 1802), to make this point and I therefore begin the chapter by outlining the strong standing of natural theology in the country and how that legacy transitioned into an adaptationism that remains to this day.

The population genetics of the modern synthesis in the 1930s put genes at the centre of evolutionary explanations. The new mathematical approach of population genetics had brought together the particulate inheritance of Mendel with the gradualism of Darwin. J.B.S. Haldane, Sewall Wright, and R.A. Fisher in their own ways formally characterized the process of evolution as one of changes in the frequency of allelic variants. They did not use the agential language of the modern gene's-eye view, but their work provided the foundations on which this approach was built. I will argue that this was especially true of the work of Fisher and that his definition of the environment only really makes sense from a gene's-eye view.

The last area was the levels of selection debate and specifically the rejection of group selection as an explanation for social behaviour. Both Williams and Dawkins pointed directly at the popularity of group selection arguments as the stimulus to write their own books. Furthermore, Williams's insistence on parsimony, that biologists should not appeal to selection at higher levels than necessary, combined with W.D. Hamilton's theory of inclusive fitness, led to a rapid change in opinion regarding how to think about adaptation.

1.2 Adaptationism and the legacy of natural theology

The biological tradition of adaptationism owes much to the field of natural theology. As an area of inquiry, natural theology is based on the assumption that the study of the natural world allows inferences to be made about the divine world. Natural theology has a long history and has been practiced in multiple faith traditions. The English approach to natural theology that came to play such an important role in the emergence of evolutionary theory was a product of particular time and place (see Alister McGrath's *Darwinism and the Divine* (McGrath 2011) for an extensive history). It came to prominence in conjunction with another quintessential English institution, the Anglican Church, and it played a central role in the Church's efforts to smooth tensions among various Protestant factions that were at odds on both theological and political issues.

Natural theology also provided a way to institutionalize the rise of science. The 19th century Oxford mathematician Baden Powell noted that the Reformation's emphasis of removing the authority of the Church led to a 'peculiarly Protestant prejudice' of constantly seeking evidence and proof for faith (quoted in McGrath 2011, p. 56). The empirical study of nature, including natural history, neatly slipped into this niche and it allowed new scientific results to be interpreted within the framework of the Church. This is not to suggest that views on the relationship between science and faith among British Protestants were completely homogeneous, instead extensive variation existed among the various denominations, as well as emerging agnostics and atheists (Stanley 2015). Nonetheless, the Church's attitude would later on lead to the rise of the scientifically literate country vicar—a profession Darwin himself aspired to when he arrived at Cambridge as an undergraduate—who could combine his church duties with a passion for natural history.

The importance of the Anglican tradition for strengthening natural theology's bond to English biology can also be seen by contrasting the English approach to that of Protestants in other parts of Europe (Ruse 2003, 2018). Take, for example, Immanuel Kant whose Lutheran Pietist upbringing gave him a deeply rooted scepticism of reason and science as a route to salvation. That scepticism was shared among many contemporary Protestants, who considered natural theology a distraction from the strict emphasis on the direct access to God for ordinary people, not just philosophers and other learned people. Rejecting natural theology did not mean Kant was sceptical of modern science. In contrast, he was deeply impressed by Newtonian physics; to him it just had nothing to contribute to matters of faith.

1.2.1 Why William Paley matters

The profile of natural theology in England was raised through a number of public events. For example, the Boyle Lectures were set up with the aim of providing a 'confutation of atheism' and 'proving the Christian Religion against notorious Infidels'. The lectures were a popular attempt by the Church of England to stem the rising tide of scepticism. No person, however, is more associated with the English approach to natural theology than the theologian William Paley. Born in Peterborough, England, in 1743 he graduated from Christ's College, Cambridge as Senior Wrangler, the university's top undergraduate student in mathematics, in 1763. He would go on to become a fellow of the college, where he gave lectures on moral philosophy. The book that emerged from these lectures, *Principles of Moral and Political Philosophy* (Paley 1785), was hugely influential. It quickly became a set text for students at Cambridge and featured in many defining debates, including in the early days of the newly formed United States Congress.

Paley's last book would become even more famous. Having already written several books on Christian apologetics, none reached the heights of *Natural Theology or, Evidences of the Existence and Attributes of the Deity, Collected from the Appearances of Nature*, published 3 years before his death (Paley 1802). It was a huge popular success and remained in print for over a century, going through over 50 editions in the UK alone. There was very little new intellectual work in *Natural Theology*. Instead, it is largely restating arguments that had been developed by others during the 17th and 18th century. There is even a good case to be made that Paley plagiarized parts of the book. In particular, Paley relied extensively on the work of the Dutchman Barnard Nieuwentyt, including the use of the watch analogy (LeMahieu 1976, pp. 60–61). Paley also ignored the serious critiques of natural theology that had been developed by then, including those by David Hume and Immanuel Kant.

Instead of intellectual novelty, it was Paley's fluent and convincing writing, coupled with a deft use of verbal imagery, that made the book stand out. The opening paragraph of *Natural Theology* is a case in point and it lays out his core argument in a few powerful sentences:

> In crossing a heath, suppose I pitched my foot against a stone and were asked how the stone came to be there, I might possibly answer that for anything I knew to the contrary it had lain there forever; nor would it, perhaps, be very easy to show the absurdity of this answer. But suppose I found a watch upon the ground, and it should be inquired how the watch happened to be in that place, I should hardly think of the answer which I had given, that for anything I knew the watch might have always been there. Yet why should not this answer serve for the watch as well as for the stone; why is it not admissible in that second case as in the first? For this reason, and for no other, namely, that when we come to inspect the watch, we perceive—what we could not discover in the stone—that its several parts are framed and put together for a purpose, e.g., that they are so formed and adjusted as to produce motion, and that motion so regulated as to point out the hour of the day; that if the different parts had been differently shaped from what they are, or placed in any other manner or in any other order than that in which they are placed, either no motion at all would have carried on in the machine, or none which would have answered the use that is now served by it.   (Paley 1802, p. 7)

Gardner (2009) describes how Paley's conception of design stands on two legs: contrivance and relation. Contrivance refers to how the constituent parts of both watches and organisms appear as if contrived for a specific purpose. Relation, in turn, captures how the parts demonstrate a unity of purpose, they are all working towards the same goal. On Paley's heath, this is what distinguishes the watch from the stone.

Just as early Anglican natural theologians were successful partly because they provided a stabilizing force in the aftermath of the English Civil War (1642–1651) and the Glorious Revolution (1688), so did Paley in his time. *Natural Theology* appeared in the midst of the European turmoil associated with the aftermath of the French Revolution. France, England's old neighbour and rival, was ruled by a threatening Napoleon and was seeing the rise of atheistic views. For example, when asked about the role of God in describing the cosmos, the astronomer Pierre-Simon Laplace is said to have replied: '*Je n'avais pas besoin de cette hypothèse-là*' (I had no need of that hypothesis). In England, the writings of Paley provided a counterweight to this attitude, reinforcing the idea that science could be done within the embrace of the established Church. In particular, he showed how the image of a mechanistic universe subject to natural laws established by Newton and his followers could strengthen belief rather than undermine it (though, Paley thought that astronomy, owing to its 'simplicity', had little to offer as evidence of the Creator; Paley 1802, p. 199).

Ultimately, *Natural Theology* was about strengthening the case for God. Given the time when it was written it was therefore no coincidence that he chose to build his case on machines and organisms. It is a book primarily about animals. Plants do get a chapter but are dismissed as unimportant: 'it is unnecessary to dwell upon a weaker argument, where a stronger is at hand', as he put it (Paley 1802, p. 183). Paley wrote in the middle of the Industrial Revolution and his readers would have been well acquainted with the watches, telescopes, stocking-mills, and steam engines that feature in his argument (Gillespie 1990). 'That an animal is a machine', Paley wrote, 'is a proposition neither correctly true nor wholly false' (Paley 1802, p. 48). Both machines and organisms are examples of contrivances and can therefore only be understood by understanding the intentions of their designer. To illustrate this, Paley provided an extensive discussion of the intricate mechanisms of a telescope in order to drill home the importance of complexity for his argument (Paley 1802, pp. 16–31).

He then points out that a human eye is far more complicated than a telescope and the eye therefore requires a more sophisticated designer.

1.2.2 Neo-Paleyan biology in Britain

Paley represented the apogee of the English approach to natural theology, especially its popular variety. Later on, the Bridgewater Treatises were commissioned by Francis Henry Egerton, 8th Earl of Bridgewater, and published from 1833 to 1840 with the goal of showing 'the Power, Wisdom, and Goodness of God, as manifested in the Creation'.

Today, few similar initiatives remain. One prominent exception is the Gifford Lectures, put on every year by the ancient Scottish universities (St Andrews, Glasgow, Aberdeen, and Edinburgh). Organized for the first time in 1888, the stated aim of the lectures is to 'promote and diffuse the study of natural theology in the widest sense of the term – in other words, the knowledge of God'. Presenting the lectures remains a prestigious honour in Scottish academia, and several distinguished contemporary theologians interested in science have given them, such as Alvin Plantinga (2004–2005) and Alister McGrath (2009). In recent years the scope of lectures seems to have strayed from the original goal. For example, several prominent atheists have also been invited, including Carl Sagan (1985), Richard Dawkins (1988), Michael Ruse (2001), Steven Pinker (2012–2013), and Sean M Carroll (2016).

The strong standing of natural theology has meant that relative to their colleagues in other countries British biologists have been preoccupied with life's design-like qualities. Already a few decades after the publication of *Natural Theology*, the collections at the Oxford University Museum of Natural History were organized by chapter orders of the book, with the stated aim of familiarizing students with Paley's argument (McGrath 2011, p. 134). Recently, this adaptationist tendency in Britain has been characterized by Tim Lewens as 'Neo-Paleyan' (Lewens 2019a).

When Darwin entered Christ's College, Cambridge, to prepare for the priesthood after dropping out of his medical studies at the University of Edinburgh, Paley was very much on the agenda. Not only because Darwin had joined Paley's old college (legend even has it that they occupied the same room), but because Paley's *Views of the Evidences of Christianity* (Paley 1794) was required reading. Alongside with *Natural Theology* it left a deep impression on Darwin who would later write:

> I am convinced that I could have written out the whole of the Evidences with perfect correctness, but not of course in the clear language of Paley. The logic of this book and, as I may add, of his Natural Theology gave me as much delight as did Euclid. (...) I did not at that time trouble myself about Paley's premises; and taking these on trust, I was charmed and convinced by the long line of argumentation.   (Darwin 1887, p. 47)

Similarly, short after the publication of the *Origin*, Darwin wrote in a letter to his friend John Lubbock: 'I do not think I ever admired a book more than Paley's Natural Theology: I could almost formerly have said it by heart' (Darwin Correspondence Project 2020a). In addition to quotes like these, the influence of Paley also shines through in the way Darwin constructs his arguments, as both Gould (2002, pp. 118–121) and McGrath (2011, pp. 156–157), have highlighted. Most importantly, both Paley and Darwin start with a human object or activity that would be familiar to the reader: a watch or telescope for Paley and breeding of farm animals in the case of Darwin. They then proceed to point out the similarities to their own proposed mechanism. Both also spend considerable time on the human eye, either to highlight its extrinsic contrivance or to demonstrate how a gradual selective process can produce the same result. In his influential history of biology, *The Growth of Biological Though*, Ernst Mayr argues that the fact that most biologists of Darwin's time were natural theologians, in practice studying what we today would refer to as adaptations, meant that most of the natural theology literature could be folded into the new field of evolutionary biology in a rather frictionless way (Mayr 1982, pp. 104–105; see also Reiss 2011, pp. 122–124).

A final example of Paley's influence is the title of Darwin's book on the co-evolution between pollinators and orchids, *On the Various Contrivances by which British and Foreign Orchids are Fertilised by Insects* (Darwin 1862), published only 3 years after *The Origin of Species*. Even at that time, 'contrivance' was a term strongly associated with natural theology and with Paley. To Darwin, natural selection was clearly the answer to the problem of design, as formulated by Paley. Moreover, when Darwin submitted the book to his publisher John Murray he wrote that 'like a Bridge-water Treatise the chief object [of the book] is to show the perfection of the many contrivances of orchids' (Darwin Correspondence Project 2020b). Similarly to how Dawkins would later praise *Natural Theology* for its excellent documentation of complex adaptations (just with the wrong conclusions drawn from the data), Darwin's book was lauded in venues like the *Literary Churchman* that it could have been read as a contribution to natural theology (just with the wrong conclusion drawn from the data). The botanist Asa Gray went so far as to say that: 'if the Orchid book (with a few trifling omissions) had appeared before the "Origin", the author would have been canonized rather than anathematised by the natural theologians' (quoted in McGrath 2011, p. 157).

Darwin's attitude to Paley and natural theology was in sharp contrast to that of his biggest supporter, T.H. Huxley. While Darwin's establishment background had meant that he had been exposed to the Anglican natural theology consensus since a young age, Huxley's more modest upbringing came with no such training. Michael Ruse has speculated that this is why Huxley remained so unimpressed with the concept of adaptation (Ruse 2019a).

Someone who was impressed by adaptations was R.A. Fisher. Fisher was not only an accomplished statistician and population geneticist, but also a deeply committed Anglican. At Cambridge, he regularly gave sermons at the college chapel and wrote essays for church magazines. (He also spent many years at the godless University College

London, which to this day lacks a divinity department.) In a letter to Darwin's granddaughter, the botanist and geneticist Nora Barlow, he admits feeling guilty for never actually finishing *Natural Theology* (Bennett 1983, p. 79). Nevertheless, Fisher considered adaptation to be the defining feature of the living world and evidence of God's continued involvement:

> To the traditionally religious man, the essential novelty introduced by the theory of evolution of organic life, is that creation was not all finished a long time ago, but is still in progress, in the midst of its incredible duration. In the language of Genesis we are living in the sixth day, probably rather early in the morning and the divine artist have not yet stood back from this work and declared it to be 'very good'.
>
> (Fisher 1947)

Paley shines through also in the writings of the next generation of British evolutionary biologists. They may not have read Paley directly but that did not prevent them from invoking his name for rhetorical purposes. One who did read Paley was Vero Copner Wynne-Edwards, author of *Animal Dispersion in Relation to Social Behaviour* (Wynne-Edwards 1962) and often assigned the role of representative of naive group selection in the current literature. Paley is said to have influenced Wynne-Edwards's concept of the 'balance of nature' and his ideas about how animals can regulate their own population size towards an optimum number that prevents the overexploitation of their resources (Borello 2010, p. 42).

The influence of Paley is even more vividly demonstrated by the attitude of two of modern British evolutionary biology's most eloquent spokesmen, John Maynard Smith and Richard Dawkins. Maynard Smith credited reading Darwin for his loss of faith and he had no hesitation regarding what he considered the central problem of evolutionary biology to be: 'the main task of any theory of evolution is to explain adaptive complexity, i.e. to explain the same set of facts which the eighteenth-century theologian Paley used as evidence of a Creator' (Maynard Smith 1969, p.82). This opinion is very much shared by Dawkins. He has described the zoology department at

Oxford that he joined as a student a strongly adaptationist (Dawkins 2015a, pp. 342–345), much thanks to the domineering presence of E.B. Ford, author of the influential *Ecological Genetics* (Ford 1964). In contrast, continental Europeans and Americans have considered the origin and constraint on diversity the central problem for evolution to explain (Sterelny 2001, pp. 4–6; Lewens 2004, p. 21). This is of course somewhat of an exaggeration, but much has been written about this strong standing of adaptationism in the UK, most extensively by Marek Kohn in *A Reason For Everything* (Kohn 2004).

Given that adaptationism, especially as represented by the Ford legacy at Oxford, was dismissed by Richard Lewontin as stemming from the 'fascination with birds and gardens, butterflies and snails that was characteristic of the pre-war upper middle class from which so many British scientists came' (Lewontin 1972), it is ironic that for long its best articulation was Arthur Cain's 1964 paper 'The perfection of animals' (Cain 1964). Cain's humble upbringing background was far from upper-middle class. At the 1979 Royal Society meeting on 'The Evolution of Adaptation by Natural Selection' organized by John Maynard Smith, Cain gave one of the responses to Stephen Jay Gould's talk (a talk that would eventually become the famous Spandrels of San Marco paper; Gould and Lewontin 1979), in which he remarked that: 'Presumably when prejudice is strong, facts can be dispensed with as well: my own background and upbringing could only be distinguished by the extreme purist from working class' (retold in Dawkins 2015a, p. 344).

After admitting that other candidates, such as the species problem or the origin of biological diversity left him cold, Dawkins argues that adaptive complexity is a unique kind of problem that deserves a unique kind of scientific explanation: either a Designer, as invoked by Paley, or something like natural selection (Dawkins 1998a). By now, Dawkins may be as famous for his public defence of atheism, as he is for his work in evolutionary biology (Giberson and Artigas 2007, pp. 19–52). His attitude to the Scottish philosopher David Hume is therefore quite informative and reveals the close bond he feels with Paley. Hume's *Dialogues Concerning Natural Religion*, published anonymously 3 years after his death, is often considered to be one of the best refutations of the design argument (Hume 1779). In the *Dialogues*, three philosophers, Philo, Demea, and Cleanthes, debate the evidence for the existence of God, in particular the argument from design. At one point, Philo, who is believed to represent Hume's own view, argues that the evidence of design can be used to support a wide range of mutually exclusive conclusions. Even that:

> This world was only the first rough attempt of some infant god, who afterwards abandoned it, ashamed of his poor performance; it is the work of some dependent, inferior god, whose superiors hold it up for ridicule; it was produced by some god in his old age and near-senility, and ever since his death the world has continued without further guidance, activated by the first shove he gave to it and the active force that he built into it.   (Hume 1779, p. 26)

To Dawkins, this misses the point. He is deeply impressed with the evidence of design and only disagrees with Paley regarding the origin of design. In his 1986 book *The Blind Watchmaker* (which for the American market was given the subtitle 'Why the Evidence of Evolution Reveals a Universe without Design'). Dawkins describe Paley as having 'a proper reverence for the complexity of the living world, and he saw that it demands *a very special kind* of explanation' (Dawkins 1986a, p. 4; my emphasis). It is this belief that led Dawkins to describe himself as a 'neo-Paleyist obsessed with the illusion of purpose, adaptation' (Dawkins 1994a; 'Paley, of course, was a paleo-Paleyist', he adds) or 'transformed Paleyist' (Dawkins 1998a, p. 16). It is also why Dawkins argues that it was only after Darwin, that it became possible to be an intellectually fulfilled atheist (Dawkins 1986a, p. 6). His point was that the weakness of the design argument had been noted before, but in the absence of a credible alternative the design argument was never mortally wounded.

As documented by Lewens (2019a), the rhetoric of Paley remains also in contemporary British evolutionary biology. The best representatives of this habit are the theoreticians Andy Gardner and Alan

Grafen. In a 2009 paper entitled 'Adaptation as organism design', which on his website he describe as his 'manifesto', Gardner writes, explicitly acknowledging, Paley's contributions: 'The problem of adaptation is to explain the apparent design of organisms. Darwin solved this problem with the theory of natural selection' (Gardner 2009). Similar language is used by Grafen when describing the rationale for his Formal Darwinism Project: 'Adaptation is the centre of biology, adaptation is design, and maximising fitness is what organisms are designed for' (Grafen 2007). In both cases, Gardner and Grafen do not shy away from the word design but instead fully embrace it as integral to what adaptation is. The issue whether evolutionary biology actually needs a design principle—an answer to the question of what organisms should appear designed to maximize—comes up in the current debate over inclusive fitness and I will return to that issue when discussing the relationship between the gene's-eye view and inclusive fitness, including the Formal Darwinism Project, in Chapter 4. For now, it suffices to say that the problem of design occupies a special place in British evolutionary biology, a place that it does not occupy in other national traditions.

It is worth noting a clear exception to this generalization, the American George Williams. Early on in *Adaptation and Natural Selection*, Williams describe adaptation as a 'special and onerous concept that should not be used unnecessarily, and an effect should not be called a function unless it is clearly produced by design and not by chance' (Williams 1966, p. 5) and then goes on to state that the book is 'an attack on what I consider unwarranted uses of the concept of adaptation' (Williams 1966, p. 11; see also Williams 1985). Like Dawkins and Maynard Smith, Williams thought that adaptation was a unique problem, and in the last chapter of the book he extensively quotes Paley's comments about the human eye to justify this. He even introduces the term teleonomy (cf. Pittendrigh 1958) as the term to describe the study of adaptation, a term subsequently endorsed by Dawkins (1982a, p. 81). Williams also included excerpts from chapters 1 and 3 of Paley's *Natural Theology* as an appendix in his last book *Natural Selection: Domains, Levels and Challenges* (Williams 1992).

Taken together, the gene's-eye view owes to natural theology in general, and to Paley in particular, its view of adaptive complexity, the appearance of design in the living world, as *the* problem that a theory of evolution needs to be able to address.

1.3 Population genetics

William Provine's landmark history of the field describes theoretical population genetics as the product of a clash of two different perspectives on evolution (Provine 1971). In one corner was Darwin's original formulation, which emphasized how natural selection acted on small continuous variations leading to gradual evolutionary change. In the other were those who thought that selection on small variants would not be enough to generate large-scale evolutionary change. This conflict intensified with the rediscovery of Mendel's work in 1900. Darwin's ideas were especially vulnerable to the attack that the Mendelian concept of particulate genetic changes was incompatible with his own theory, as he had lacked a functioning theory of inheritance of his own. The formulation of population genetics resolved the disagreement. The key works were all published in an intense decade and a half between 1918 and 1932 by the Britons R.A. Fisher and J.B.S. Haldane, and the American Sewall Wright.

Fisher, Haldane, and Wright all contributed both to the reconciliation of Darwin and Mendel and to the introduction of theoretical population genetics as it is known it today. They all championed a new mathematical way of doing biology, an approach that was not appreciated by everyone. Ernst Mayr, for example, famously questioned the value of their contributions: 'What, precisely, has been the contribution of the [Fisher, Wright, and Haldane] mathematical school to evolutionary thinking? If I may be permitted to ask such a provocative question' (Mayr 1959). Mayr coined the term 'beanbag genetics' to describe their approach, resulting in Haldane's spirited

'A defense of beanbag genetics' the following year (Haldane 1964). The exact views of Fisher, Wright, and Haldane, however, differed quite markedly. Provine described the relationship between their views as 'symmetrical', in the sense that each of them thought the other two had more in common (Provine 1971, p. 176). They disagreed on issues such as the importance of dominance, epistasis, drift, and whether selection would be more efficient in large or small populations. Who of the three got most things right is a popular debate at happy hour among graduate students in evolutionary biology and in the more sober reflections of professional historians of science. I will argue, however, that if we restrict ourselves to the gene's-eye view, Fisher towers above the other two.

The intellectual foundations of the gene's-eye view can in retrospect be said to be implicit in the writings of Haldane and Wright, but to be explicit in the writings of Fisher (Sarkar 1994; Okasha 2008a; Ewens 2011; Edwards 2014). I will therefore focus on the views of Fisher and especially on how his major contribution, the Fundamental Theorem of Natural Selection, opened up the avenue to think in terms of a gene's-eye view.

1.3.1 Ronald Aylmer Fisher

The mathematical talents of R.A. Fisher were recognized early. Already as a seven-year-old he attended lectures by the Irish astronomer and celebrated popularizer of science Sir Robert Ball and these talents would later be nurtured as a boarder at Harrow School. Fisher famously suffered from a greatly reduced eyesight, which meant that he was not allowed to work under direct electrical light. His tutors at Harrow therefore had to teach him mathematics without visual aids and without pen and paper. It is said that this left him with an unusual mathematical intuition and an ability to solve mathematical problems in his head. This ability has also been suggested to be the reason why many of us with lesser mathematical minds struggle with the many mathematical intuitive leaps that characterize Fisher's writing. Indeed, as pointed out by Alan Grafen: 'if your understanding of [Fisher's Fundamental Theorem of Natural Selection] comes from reading Fisher himself, you should be aware that virtually no other biologists have understood what he meant' (Grafen 2003). Fisher left Harrow for Gonville and Caius College, Cambridge in 1909. Like Paley, he was a Wrangler in the Mathematical Tripos, achieving the honour in 1912. His fascination with biology began before arriving in Cambridge. As reported by his daughter Joan Fisher Box, he brought with him to Cambridge a thirteen-volume set of the complete works of Darwin (a school prize; Fisher Box 178, p. 17). While there, his interest in evolution was further spurred by reading Karl Pearson's collection of papers *Mathematical Contributions to the Theory of Evolution*. Pearson was from the generation before Fisher's and his interest in mathematical statistics, of which he is widely considered one of the founding fathers, and evolutionary theory in the form of biometry (as well as eugenics) meant that he left a strong impression on the young Fisher. Later, the two of them would both have offices in the same building at the University College London. Personal and scientific disagreements, however, would eventually lead to the two of them falling out.

1.3.2 Fisher (1918) and the birth of the gene's-eye view

If you were to put a date on the origin of the gene's-eye view, Fisher's graduate student A.W.F. Edwards has suggested 1 October 1918 (Edwards 2014). This was the publication date of Fisher's 'The correlation between relatives on the supposition of Mendelian inheritance' in the *Transactions of the Royal Society of Edinburgh*. Fisher had initially submitted the paper to the Royal Society of London, but unfavourable reviews from among others R.C. Punnett (of Punnett squares) and Karl Pearson prevented publication. The machinations behind the rejection also included G.H. Hardy (of the Hardy–Weinberg equilibrium; see Norton and Pearson 1976 for the full story). Like much of Fisher's writing, the 1918 paper is a tough read, but it is worth the slog. As suggested by the title, Fisher's goal is to examine correlations in traits between relatives in light of Mendelian inheritance (a topic previously addressed by Pearson who reached different conclusions, hence his negative reaction).

The contributions of the paper were many (Moran and Smith 1966). In particular, Fisher calculated the expected values of the correlation between relatives under a variety of scenarios, including linkage, partial recessivity, epistasis, and assortative mating. To do this, Fisher developed the statistical method of regressing genotypic value on the number of alleles at a given locus (0, 1, or 2 in a diploid), which also led to the introduction of the term variance.

The concept of variance is key to understanding why the 1918 paper can be thought to be the birth of the gene's-eye view. Edwards (2014) explains it as follows. Consider a locus with genotypes AA, Aa, and aa. Their genotypic values are i, j, and k and are present in the population with frequencies P, $2Q$, and R respectively. Now, using a least-square regression approach we can calculate the regression coefficient, α, of the genotypic value given the number of A alleles, such that:

$$\alpha = \frac{P(Q+R)(i-j) + R(P+Q)(j-k)}{P(Q+R) + R(P+Q)}$$

The regression coefficient, α, is the expected change in the genotypic value for a given substitution, i.e. going from aa to Aa, or Aa to AA. With this in place, Fisher's genetic variance, the part that he calls the 'additive part which reflects the genetic nature without distortion', can be written as:

$$(P+R)\alpha^2 - (P\alpha - R\alpha)^2$$

Fisher contrasts this additive genetic part with a residue part 'which acts in much the same way as an arbitrary error introduced into the measurements'. This, then, is the crux: any variation not included in the variance captured by the linear regression should not be considered genetic but instead be thought of as random environmental noise.

This may seem like a subtle point, but here Fisher introduces a radically expanded notion of the environment. The expanded notion of the environment is awkward from the perspective of the individual organism, but makes sense under a gene's-eye view. From a gene's-eye view, or more precisely from the perspective of an allele, Fisher's notion has the consequence that in a diploid organism, the other allele present at the same location in the genome should be considered part of the environment. And so should all other genes in the genome, as well as all the other genes in the population (the gene pool). This idea of the genomic environment is key to appreciating the deep connection between Fisher's Fundamental Theorem and the gene's-eye view. It also highlights how the gene's-eye view could more accurately be called the allele's-eye view.

Fisher expanded on these ideas in his 1922 paper 'On the dominance ratio', also in a journal run by the Royal Society of Edinburgh (Fisher 1922). This paper included, among other things, the first demonstration of the consequences of heterozygote advantage, a phenomenon that later would be used by Elliott Sober and Richard Lewontin as an argument against the gene's-eye view. I will return to this topic in Chapter 3. In the succeeding years, Fisher focused on applying his theoretical models to empirical data and worked on issues like genetic variation in British moths (Fisher and Ford 1926) and mimicry (Fisher 1927). In 1930, he synthesized his ideas his most famous publication, *The Genetical Theory of Natural Selection* (Fisher 1930).

1.3.3 *The Genetical Theory of Natural Selection*

Julian Huxley called *The Genetical Theory of Natural Selection* 'the most important book to have come out this century' (quoted on the jacket of the 1999 Oxford University Press Variorum Edition). Almost a century later many would still agree and the book ranks among the most important of the modern synthesis. Other than the mathematical sections, the manuscript was typed by his wife, Eileen, mostly in the evenings and weekends, as Fisher spent his weekdays as Head of the Statistics Department at Rothamsted Experimental Station. In retrospect, Fisher would write that the chief goal of the book was to confront the claim that 'the discovery of Mendel's laws of inheritance was unfavourable, or even fatal, to the theory of natural selection' (Fisher 1954). Spanning twelve chapters it opens with a Preface that begins with the famous sentence: 'Natural Selection is not Evolution'. Arguably the most important part of the book, at least for our purposes here, is Chapter 2 where he introduces the Fundamental Theorem of Natural Selection.

If *The Genetical Theory* is still highly regarded, the Fundamental Theorem of Natural Selection, has a more mixed legacy. That Fisher himself thought highly of the Fundamental Theorem is implied not only by its grand name but also by his likening it to the second law of thermodynamics (Fisher 1930, p. 36) and his description of it as occupying 'the supreme position among the biological sciences' (Fisher 1930, p. 37). In contrast, the attitudes of most evolutionary biologists of subsequent generations have been lukewarm, if not directly hostile. The mathematician Sam Karlin is often attributed the comment (though usually without reference, for example, Walsh and Lynch 2018, p. 154) that the Fundamental Theorem is 'neither fundamental nor a theorem', which was the consensus position among theoretical population geneticists up until at least the mid 1990s. At that time, a number of authors used George Price's 1972 paper to reinterpret the Fundamental Theorem, reinstating some of the importance Fisher had awarded it (Price 1972; Edwards 1994; Grafen 2003; Queller 2017). The disagreements between what has been called the traditional and modern interpretations of the Fundamental Theorem continue to this day (see, e.g., the exchange between Grafen 2018 and Lessard and Ewens 2019 for contrasting views).

Here, I am less concerned with taking sides about the mathematical correctness or biological significance of the Fundamental Theorem, but rather with evaluating its connection to the gene's-eye view. To do so, however, requires us to unpack the assumptions of the theorem and to derive the origin of the traditional and modern interpretations.

1.3.4 The gene's-eye view and the Fundamental Theorem made clear

Early on in *The Genetical Theory,* Fisher defined the Fundamental Theorem as: 'The rate of increase in fitness of any organism at any time is equal to its genetic variance in fitness at that time' (Fisher 1930, p. 35). One look at this original definition quickly provides some insight why it has been marred by misunderstandings and disagreements. What does genetic variation in fitness of an organism even mean? Later on in Chapter 2, Fisher used 'species' instead of 'organism' and in 1941 replaced it with 'population'. In 1941, he also added the qualification 'ascribable to a change in gene frequency'. Both changes help clarify that Fisher's central claim is that the rate at which natural selection increases the average fitness of a population through changes in gene frequencies will be equal to the genetic variation in fitness at the time (Box 1.1).

The traditional attitude that the Fundamental Theorem mostly fails lasted until 1989 when A. W. F. Edwards published a paper in *Theoretical Population Biology* bringing attention to a long-neglected 1972 paper by the brilliant and troubled American George Price (see Harman 2011 for a gripping biography). Price's 'Fisher's "fundamental theorem" made clear' spans only twelve pages, but its influence on changing attitudes towards the Fundamental Theorem was immense. Grafen (2003) described it as the 'path of enlightenment towards the fundamental theorem'. As discussed above, no one should feel bad if they struggle to understand Fisher. Price put it thus: 'Fisher's explanations of his theorem are afflicted by a truly astonishing number of obscurities, infelicities of expression, typographical errors, omissions of crucial explanations, and contradictions between different passages about the same point' (Price 1972). The key point that Price made clear about the Fundamental Theorem was that what Fisher was

Box 1.1. Unpacking the Fundamental Theorem

The Fundamental Theorem can be divided into two key concepts (Okasha 2008a). The first is simply the average fitness of a population. If we define fitness of the ith individual as ω_i then we can write the average fitness of a population of size n as $\bar{\omega} = \dfrac{\sum_i^n \omega_i}{n}$. The change in the average fitness over time can then written either as $\dfrac{d\bar{\omega}}{dt}$ in a continuous model (as Fisher originally did) or, following Ewens (1989), as $\Delta\bar{\omega}$ for a discrete time model (the discrete time steps can, e.g., be from one generation to the next).

The second concept, the genetic variance in fitness, takes a little longer to unpack. I will largely follow Okasha's (2008a) terminology, including his slightly simplified version of what Fisher called the 'average effect' further down.

Fisher's unusual mathematical intuitions have made many of us struggle with his arguments. It is well established that variation in fitness can be due to genetic differences among individuals, environmental variation, or a combination of the two. The way Fisher talks about 'genetic variance' can therefore be confusing. He does not refer to the total variation in fitness due to differences in genotype between individuals, $Var(g)$, but instead to what we now typically refer to as additive genetic variance. Additive genetic variance, $Var_{add}(g)$, is only part of the total genetic variance, with the non-additive (sometimes called epistatic, which also includes dominance) variance, $Var_{non-add}(g)$, making up the other part, such that:

$$Var(g) = Var_{add}(g) + Var_{non-add}(g)$$

Additive genetic variance (Fisher's 'genetic variance') captures how a gene affects fitness, independently of what other genes are present in the genome. To see what this means, imagine the extreme case where there are no interactions between genes and that they all act completely independently of each other. In this scenario, the total genetic variance will equal the additive variance, that is $Var(g) = Var_{add}(g)$. Now, using Fisher's (1918) regression method described above, the fitness of a diploid individual, i, can be written as:

$$\omega_i = \sum_j \alpha_j x_{ij} + e_i$$

Recall, that α was defined earlier in the chapter as the regression coefficient of fitness on the number of copies of a given allele so that x_{ij} is the number of j alleles carried by individual i. α_j thus tells us how much the fitness of an individual changes if you were to add one more copy of the j allele (assuming the rest of the genome stays the same). This is what Fisher called the 'average effect'. e_i is the residual term and under perfect additivity it would be 0 and ω_i would equal $\sum_j \alpha_j x_{ij}$.

The additive genetic variance, $\mathrm{Var_{add}}(g)$, is then equal to:

$$\mathrm{Var} \sum_j \alpha_j x_{ij}.$$

Under the traditional interpretation of the Fundamental Theorem, Fisher's was thought to have argued that the change in the average fitness over time $\left(\dfrac{d\bar{\omega}}{dt} \right)$ would equal the additive genetic variance, $\mathrm{Var_{add}}(g)$ (or $\dfrac{\mathrm{Var_{add}}(g)}{\bar{\omega}}$ in the discrete-time model). In both cases, this seems to mean that the average fitness of a population will never decrease.

If this indeed what Fisher meant, it is easy to see why so many people have been sceptical. There are many ways fitness may decrease in a population, even if it is subject to natural selection and the change in average fitness of the population will thus not be equal to the additive genetic variance (as a variance cannot be negative).

concerned with was not the total change in average fitness of a population, but the part of the total change that is due to natural selection changing gene frequencies in a constant environment.

Price's insight about the constant environment is key. Recall that Fisher introduced a radically broad definition of environment that included all the genes in the genome, as well as the population's gene pool, when he developed the regression method to determine an allele's average effect. Fisher's idea of a constant environment meant that the average effect of all alleles stays the same from one generation to the next. Thus, if the average effect of one allele has changed, the environment is considered to have changed. Price's insight about the constant environment also reveals the deep connection between the Fundamental Theorem and the gene's-eye view. Grouping biotic environmental effects (what biologists usually mean by the word) with non-additive genetic effects, such as dominance and epistasis as parts of the environment, only really makes sense under a gene's-eye view (but see Sober 2020 for a pushback against this conclusion).

The importance of this conception of the environment runs through most Dawkins's writings. For example, in the foreword to the 30th anniversary edition of *The Selfish Gene* he writes:

> Each gene [is] pursuing its own self-interested agenda against the background of the other genes in the gene pool—the set of candidates for sexual shuffling within species. Those other genes are part of the environment in which each survives, in the same way was the weather, predators and prey, supporting vegetation and soil bacteria are part of the environment. From each gene's point of view, the 'background' genes are those with which it shares bodies in its journey down the generation. In the short term, that means the other members of the genome. In the long term, it means the other genes in the gene pool of the species. Natural selection therefore sees to it that gangs of mutually compatible—which is almost to say cooperating—genes are favoured in the presence of each other.   (Dawkins 2006a, pp. xii–xiii)

The idea of a genomic environment is even more clearly laid out in the chapter titled 'The selfish cooperator' in *Unweaving the Rainbow* where he writes:

> At each genetic locus, the gene most likely to be favoured is the one that is compatible with the genetic climate afforded by the others, the one that survives in that climate through repeated generations. Since this applies to each one of the genes that constitute the climate—since

every gene is potentially part of the climate of every other—the result is that a species gene pool tends to coalesce into a gang of mutually compatible partners.   (Dawkins 1998b, p. 215)

In both quotes, the influence of Fisher shines through. Indeed, Michael Wade would later criticize *The Selfish Gene* by saying that Dawkins's argument would have worked '[i]f evolution in natural populations followed the paradigm developed by R. A. Fisher' (Wade 1978). In his original formulation, Dawkins made this argument using the analogy of a rowing crew and I will return to this when discussing epistasis and the so-called averaging fallacy in Chapter 3.

Finally, whether we accept Fisher's expanded notion of environment will also determine our attitude to the biological significance of the Fundamental Theorem. Although, Price (1972) was instrumental in shifting attitudes towards the mathematical correctness of the Theorem, he described Fisher's idea of treating non-additive gene effects as environment as a 'defect' of the Theorem (Price 1972, p. 139). According to Price, this defect meant that the environment is always changing and that we can therefore not actually say anything about whether mean fitness will increase or decrease. The Theorem can only make such a prediction in a constant environment. On the other hand, from a gene's-eye view Fisher's expanded environment is not a mere mathematical trick, with no biological or theoretical justification, but the logical consequence of the shift of perspective from organism to gene. Taking a gene's-eye view therefore makes it easier to accept the biological significance of the Fundamental Theorem.

1.4 Levels of selection

The third strand of evolutionary thought that the gene's-eye view owes its existence to is the levels of selection debate of the 1960s and 1970s. Though once dismissed as a 'rather foolish controversy' (Waddington 1975), disagreements over which level in the biological hierarchy—genes, organisms, groups, species—natural selection occurs later developed into one of the liveliest areas of evolutionary biology and indeed 'one of the brightest areas in recent philosophy of science' (Dennett 1995, p. 327). An important reason for the longevity of the issue is that the debate involves both conceptual and empirical elements.

Group selection is the idea that group-level adaptations are the product of selection not on individuals but on groups. Because group selection can act in the opposite direction of selection at the individual level, it can provide a mechanism for the evolution of altruism, social behaviours that come with a fitness cost to the individual performing them but with a benefit to the recipient and/or the group as a whole. While theoreticians today agree that formal and proper models of group selection can be formulated, the situation in the past century was rather different, and talk about how individuals behaved 'for the good of the group' and in order to preserve the existence of the species was rampant.

The contentious history of group selection has been extensively reviewed, and I will therefore provide only a brief sketch here. Anyone looking for a longer version has several books to explore. Do note, however, that they often arrive at diametrically different conclusions regarding the value of group selection (it has even been difficult to agree whether Darwin himself used group selection arguments; Ruse 1980; Borello 2005; Gardner 2011). To get a sense of the full spectrum of opinion on group selection, compare the very critical *The Ant and the Peacock* (Cronin 1991) and the much more positive *Unto Others* (Sober and Wilson 1998) or *Evolutionary Restraints* (Borello 2010). More recently, Michael Wade has provided a lovely personal account of much of this history in *Adaptation in Metapopulations* (Wade 2016). Several anthologies with key papers of the general levels of selection debate exist (examples include Brandon and Burian 1984 and Keller 1999). Anyone looking for a guide to the many conceptual issues related to group selection controversy, and the broader levels of selection debate, can do no better than to consult Samir Okasha's *Evolution and the Levels of Selection* (Okasha 2006).

1.4.1 Wynne-Edwards and the origins of naive group selection

Borello (2005) has characterized the first half of the 20th century as a 'period of mutual acceptance, or at least benign neglect' when it came to the group selection question. Fisher, Haldane, and Wright all touched on it in one way or another in their writings. Fisher allowed it some role in the maintenance of sexual reproduction but also argued that the fact that individuals reproduce faster than groups means that selection on individuals will usually overpower that on groups (Fisher 1930, pp. 122–123). Furthermore, in an essay titled 'Some hopes of a eugenicist', he also outlined a kin selection argument for how nephews may 'genetically' replace a childless uncle (Fisher 1914). Haldane is famous for the quip that 'he was prepared to lay down his life for eight cousins or two brothers', a story recounted by John Maynard Smith in his review of E.O. Wilson's *Sociobiology* (Maynard Smith 1975), an idea that he explored in Haldane (1955). In *The Causes of Evolution,* Haldane also discussed examples of traits that are advantageous for a group but harmful for the individual (Haldane 1932, p. 119). Wright appears to have been most sympathetic towards group selection and a version of it can be said to be part of his shifting balance theory (Wright 1931). However, he did not discuss it in the context of social behaviour. As usual there was thus some variation in their views, but group selection was not a key point of contention among them.

The period of mutual acceptance came to an abrupt end in 1962. The publication of Vero Copner Wynne-Edwards's *Animal Dispersion in Relation to Social Behaviour* kicked off the group selection debate as it is known it today. Wynne-Edwards was not the first person to argue in favour of group selection. Consider, for example, this section from the conclusion of the influential 1949 textbook *Principles of Animal Ecology:*

> The probability of survival of individual living things, or of populations, increases with the degree to which they harmonically adjust themselves to each other and the environment. This principle is basic to the concept of the balance of nature, orders the subject of matter of

ecology and evolution, underlies organismic and developmental biology, and is the foundation for all sociology.   (Allee et al. 1949, p. 729)

Nevertheless, Wynne-Edwards, who read zoology at New College, Oxford, where he was briefly tutored by Julian Huxley, has been the name most strongly associated with group selection. Across 23 chapters and 653 pages, *Animal Dispersion* is one long argument that selection on individuals is not enough to explain social behaviour. In particular, Wynne-Edwards was concerned with how animals are prevented from overexploiting their resources—what he referred to as 'the balance of nature'. He argued that animals had evolved the ability to infer the size of their population and then regulate it accordingly to avoid the population crashing from lack of resources. Today, this is a position that even the strongest proponents of modern group selection theory call 'naive' (e.g. Wilson and Wilson 2007).

Wynne-Edwards's proposal would set off a flurry of counterarguments. One of the first salvos came from Wynne-Edwards's nemesis David Lack. Lack had published *The Natural Regulation of Animal Numbers* in 1954, a book that had touched on similar topics as *Animal Dispersion* (Lack 1954). Lack came down heavily on side of individual selection, which had prompted Wynne-Edwards to write his own book. In his 1966 *Population Studies of Birds,* Lack reiterated many of the same arguments before dedicating a full appendix just to Wynne-Edward's book (Lack 1966). The same year another heavy blow to the reputation of group selection would land: the publication of *Adaptation and Natural Selection* by George C. Williams.

1.4.2 George Christopher Williams and *Adaptation and Natural Selection*

George Christopher Williams was born in Charlotte, North Carolina, United States in 1926 and grew up in New York and Maryland. He did military service in Italy during World War II, working on water purification. He subsequently enrolled at University of California at

Berkeley, where he was taught by the botanist and modern synthesis architect G. Ledyard Stebbins. In 1955, he obtained his PhD from the University of California, Los Angeles. Following appointments at the University of Chicago and Michigan State University, he was recruited to the State University of New York at Stony Brook, where he spent the remainder of his career, playing a key role in building up the department to be one of premier centres for evolutionary biology in the world.

In addition to his work on the levels of selection, he made numerous contributions to evolutionary theory (Futuyma and Stearns 2010; Stearns 2011). For example, his 1957 *Evolution* paper 'Pleiotropy, natural selection, and the evolution of senescence' introduced the idea of antagonistic pleiotropy, where an allele has a positive fitness effect early in the life cycle and a negative one later on, as an explanation for ageing (Williams 1957). Together with Randolph Nesse, he also founded the field of evolutionary medicine (Nesse and Williams 1994). In 1999, he shared the Crafoord Prize in Biological Sciences from the Royal Swedish Academy of Sciences with Ernst Mayr and John Maynard Smith for their 'pioneering contributions to broadening, deepening and refining our understanding of biological evolution and related phenomena'. He passed away in 2010 following a period of Alzheimer's disease.

Alongside *The Selfish Gene,* Williams's *Adaptation and Natural Selection* is the defining text of the gene's-eye view. Williams's book was directed towards professional biologists rather than the general public and is therefore less known than *The Selfish Gene.* Its influence on the academic field of evolutionary biology, however, has been enormous and it remains lauded by both critics (e.g. Sober and Wilson 2011) and admirers (e.g. Boomsma 2016).

In the preface of the 30th anniversary edition of the book, Williams recounts his motivations for writing it (Williams 1996a, p. ix). He was spending the 1954–55 academic year at the University of Chicago on a teaching fellowship. Chicago at the time was the home of people like W.C. Allee and A.E. Emerson and was a stronghold for group selection (Borello 2010). One particular day, Emerson gave a lecture on what he had dubbed 'beneficial death', that individuals would sacrifice their own lives for the benefit of the group. So distraught with the message of the talk, Williams told his wife, the biologist Doris Williams, that if such a presentation was 'acceptable biology, [he] would prefer another calling'. Later on, Williams would even go so far as to say that if what Emerson was talking was consider sound biology, he would be better off selling insurance (Williams 1996b). Similarly, Dawkins has also singled out popular books on group selection, such as Konrad Lorenz's *On Aggression* (Lorenz 1963 in original German, English translation 1966) and Robert Ardrey's *The Social Contract* (Ardrey 1970), as providing the impetus to write his own book: 'to undo the damage done by Ardrey and Lorenz—and by many television documentaries of the time' (Dawkins 2013a, p. 261).

1.4.3 Three mistakes of naive group selection

Bentley (2020) has identified three key assumptions that the early naive group selectionists were mistaken about. The first assumption was that group selection was required for social behaviour that comes with a cost to the individual performing it to evolve. As Wynne-Edwards put it:

> when the short-term advantage of the individual undermines the future safety of the race, group-selection is bound to win, because the race will suffer and decline, and be supplanted by another in which antisocial advancement of the individual is more rigidly inhibited.
>
> (Wynne-Edwards 1962, p. 20)

W.D. Hamilton showed that the idea that if selection at the individual level was favouring selfish behaviour, group selection will overpower it was false (Hamilton 1963, 1964a, 1964b). I will explore the details of Hamilton's ideas regarding social evolution, including the modern understanding of inclusive fitness and Hamilton's Rule and how they relate to the gene's-eye view, in Chapter 4. For now, it is enough to note that at he demonstrated that costly social behaviours can evolve by individual level selection if it is preferentially directed towards relatives.

The second assumption was that group selection as Wynne-Edwards imagined it would be a strong selective force. In 1963, *Nature* published a brief note by Wynne-Edwards, very much a précis of the book (Wynne-Edwards 1963), which prompted John Maynard Smith to respond in this same venue (Maynard Smith 1964). Maynard Smith's reply is memorable for two reasons. To start, relying on Hamilton (1963), he coined the term kin selection to describe Hamilton's insight. Next, he introduced his own very influential haystack model of group selection. In his model, Maynard Smith imagines a species of mouse that lives in haystacks spread across a meadow. Each year, mice colonize each haystack. Once there, the mice and their offspring reproduce throughout the summer until the harvest begins and the mice disperse into the meadow. Next year, new haystacks are colonized by mice that survived the previous year. The mice come in two phenotypes: timid and aggressive. The aggressive mice outcompete the timid mice within a haystack, resulting in only aggressive individuals present at the end of the summer. However, a haystack made up of only timid individuals will produce more individuals than do haystacks with aggressive mice present. While this difference provides a way for group selection to operate, Maynard Smith showed that when there is migration between the haystacks, this will quickly lead to a scenario where the aggressive mice win out. The main take-home message of the haystack model is therefore that group selection will usually be a weak force in nature. It is not 'bound to win' as Wynne-Edwards had put it (Wynne-Edwards 1962, p. 18).

The final mistaken assumption was that group adaptations were common in nature. This is one that Williams forcefully went after in *Adaptation and Natural Selection*. Rather being the product of group selection, he argued, many, if not most, putative examples of group adaptation were actually examples of individual level adaptations. The subtitle of *Adaptation and Natural Selection* was 'A critique of some current evolutionary thought' and early on Williams describes the book as 'an attack on what I consider unwarranted uses of the concept of adaptation' (Williams 1966, p. 11). As discussed above, Williams thought of adaptation as 'a special and onerous concept' and, to him, the most important function of the book was to call for the development of 'an effective set of principles for dealing with the general phenomenon of biological adaptation' (Williams 1966, p. 19). The book was relentless in its demands for clarity regarding concepts like adaptation, selection, and fitness and it raised the standard of argument in evolutionary biology considerably.

Williams took particular aim at the confusion between group adaptations and 'fortuitous group benefits'. On this, Williams's writing is so clear that little can be added:

> Benefits to groups can arise as statistical summations of the effects of individual adaptations. When a deer successfully escapes from a bear by running away, we can attribute its success to a long ancestral period of selection for fleetness. Its fleetness is responsible for its having a low probability of death from bear attack. The same factor repeated again and again in the herd means not only that it is a herd of fleet deer, but also that it is a fleet herd. The group therefore has a low rate of mortality from bear attack. When every individual in the herd flees from a bear, the result is effective protection of the herd.   (Williams 1966, p. 16)

In other words, a herd of fleet deer is not the same as a fleet herd of deer. Just because a trait is good for the group, it does not mean it evolved by group selection. Selection on each individual deer for its running speed may lead to the fortuitous benefit of a fast running group. In general, Williams argued that, in the name of parsimony, biologists should not appeal to selection at higher levels than necessary.

Williams had an incredible influence on biologists' attitudes towards group selection. He noted that while he was confident that his views would eventually be considered orthodox, he was surprised how quickly they became so. Indeed, the shift in attitude among evolutionary biologists towards group selection turned out to be unusually rapid. Geoffrey Parker would later say that in order to get published in 1965 you had to be a group selectionist, whereas 10 years later you had to be a kin selectionist (quoted in Segerstrale 2000, p. 54).

In general, the study of social behaviour in the 1960s and 1970s was buzzing with new theoretical advances. In addition to Hamilton, Maynard Smith, and Williams, Robert Trivers had a large influence through his papers on reciprocal altruism (Trivers 1971) and on parent–offspring conflict (Trivers 1974). R.A. Fisher (Fisher 1958) and Richard Lewontin (Lewontin 1961) had introduced game theory as a theoretical framework to evolutionary biology, and this approach reached a larger audience with the publication of Maynard Smith and Price's 'The logic of animal conflict' in *Nature* in (Maynard Smith and Price 1973). Finally, in 1975 E.O. Wilson published *Sociobiology*, which was a colossal documentation of social behaviour among animals including humans (Wilson 1975), which set off the contentious sociobiology debate about the place of evolutionary models in the study of human behaviour.

1.4.4 Calling genes selfish

Clinton Richard Dawkins was born in 1941 in Nairobi, in what was then The Colony and Protectorate of Kenya and part of the British Empire (Dawkins 2013a, 2015a). He spent his early years in southern Africa, with his family in Nyasaland (now Malawi) and as a boarder in Rhodesia (now Zimbabwe). The family returned to England when he was eight. In 1959 he enrolled at the University of Oxford, where he (with the exception of a brief stint as a faculty member at the University of California, Berkeley) would spend his whole professional career. After receiving his doctorate for a thesis on decision making in chicken supervised by the Dutch ornithologist and future Nobel Prize winner Niko Tinbergen (Dawkins 1966), he subsequently rose through the ranks at the Department of Zoology, being appointed lecturer in 1970 and reader in 1990. In 1995, he became the first Simonyi Professor for the Public Understanding of Science, a position he held until his retirement.

As the author of numerous best-selling books and presenter of award-winning documentaries on the subject of evolution, creationism, and atheism, Dawkins is a strong contender for the most famous evolutionary biologist alive. In 1966, however, when Tinbergen was leaving for a sabbatical and asked him to take over his lectures on animal behaviour, he was young, inexperienced, and very nervous. Having never lectured to large groups of undergraduates before, he took the time to type out his notes in order to help himself stay on track, in particular when navigating the tricky waters of Hamilton's recently published theory of inclusive fitness (Dawkins 2013a, p. 196). Looking at these notes today, the rhetoric that would characterize much of his subsequent writing is quickly noticeable:

> Genes are in a sense immortal. They pass through generations, reshuffling themselves each time they pass from parent to offspring. The body of an animal is but a temporary resting place for the genes; the further survival of the genes depends on the survival of that body at least until it reproduces and the genes pass into another body (...) the genes build themselves a temporary house, mortal, but efficient for as long as it needs to be (...) To use the terms 'selfish' and 'altruistic' then, our basic expectation on the basis of the orthodox neo-Darwinian theory of evolution is that *Genes will be 'selfish*.
>
> (Retold in Dawkins 2013a, p. 264; Original emphasis).

This is where the idea of 'selfish genes' is born. Despite being published in the same year as Dawkins was preparing his lectures, he would not read *Adaptation and Natural Selection* until a few years later. He immediately recognized the intellectual bond between their world views. Many of the ideas in *The Selfish Gene* are present in *Adaptation and Natural Selection*, but to Dawkins, the way Williams expressed ideas was 'too laconic, not full throated enough' (Dawkins 2016, p. xxiii).

As will be become clear in the next few chapters, however, calling genes selfish is in equal parts helping and harming in communicating that point. The term 'gene's-eye view', which I have used throughout this book, would not, to the best of my knowledge, appear until in 1980 in a paper by David P. Barash on evolutionary approaches to understanding human families (Barash 1980, p. 187). Barash coined the term borrowing from the expression 'God's-eye view' (D.P. Barash, personal communication) in an attempt to apply Hamilton's work on kin selection to humans.

1.5 Summary

- The gene's-eye view takes adaptation, the appearance of design in living world, to be the central problem that a theory of evolution needs to explain. This tradition has been particularly strong in British biology, much thanks to the strong standing of natural theology and the writings of William Paley.
- A gene's-eye view can in retrospect be said to be implicit in the work of Haldane and Wright, but it is clearly explicit in Fisher's. In particular, Fisher (1918) introduces a subtle but radical shift in what should be considered the environment and is arguably the first paper to make use of a gene's-eye view. Fisher's Fundamental Theorem of Natural Selection also makes the most sense when adopting a genic perspective.
- Group selection has a tumultuous history and the levels of selection debate provided the impetus for George Williams and Richard Dawkins to write *Adaptation and Natural Selection* and *The Selfish Gene* respectively.

2

Defining and Refining
Selfish Genes

2.1 Introduction

Around the time of its publication, *The Selfish Gene*'s editor
Michael Rodgers was discussing the book's prospects with his
colleague at Oxford University Press, Richard Charkin. The initial
print run was 5,000 copies, but Charkin thought it would struggle to
sell more than 2,000. Rodgers, the optimist, promised to pay Charkin
£1 for every 1,000 copies sold under 5,000, and Charkin was to buy
Rodgers a pint of beer for every 1,000 copies over 5,000. To date, the
book has gone through four editions, been translated into twenty-five
languages, and sold well over a million copies. As Charkin puts it, he
is 'holding back payment in the interests of [Rodgers's] health and
well-being' (Rodgers 2013, p. 53). Similarly, half a century on, *Adaptation
and Natural Selection* remains in print: 1996 saw the publication of a
30th anniversary edition with an updated preface by George Williams,
and 2018 another paperback edition with a new foreword by Richard
Dawkins.

Both *Adaptation and Natural Selection* and *The Selfish Gene* were
widely reviewed. Williams was highly praised in the scientific literature
and the book received very favourable reviews. Richard Lewontin
called it 'excellent' in *Science* (Lewontin 1966) and Lawrence Slobodkin,
though he had quibbles with some of Williams's conclusions, titled his

The Gene's-Eye View of Evolution. J. Arvid Ågren, Oxford University Press. © J. Arvid Ågren 2021.
DOI: 10.1093/oso/9780198862260.003.0003

long review in *The Quarterly Review of Biology* 'The light and the way in evolution' (Slobodkin 1966). *The Selfish Gene* amassed over 100 reviews, both in the popular press and scientific journals, most of which were positive (Dawkins 2013a, p. 281). *The New York Times,* for example, described it as 'the sort of popular science-writing that makes the reader feel like a genius' (Anonymous 1977).

In contrast to *Adaptation and Natural Selection*, *The Selfish Gene* also attracted some very strong and contrasting views. W.D. Hamilton wrote an enthusiastic review for *Science* (Hamilton 1977a) saying that the book 'should be read, can be read, by almost everyone'. Charles Langley, on the other hand, wrote a scathing review for *Bioscience*, calling it 'shallow and untrue to the science of evolutionary biology' and 'a nuisance to the knowledgeable reader and misleading to the layman' (Langley 1977). Along the same lines, Richard Lewontin's review for *Nature* (Lewontin 1977a) was entitled 'Caricature of Darwinism' and called Dawkins's adaptationist thesis 'Panglossian' (2 years before the Spandrels of San Marco paper with Stephen Jay Gould), and also singled out the journal the *American Naturalist* as an especially egregious home of this habit. The tone of Lewontin's writing led Hamilton to pen a letter of protest to the editor, where he called the review a 'disgrace and compared it to Bishop Wilberforce's attack on Darwin and Huxley at the British Association meeting in 1860 (Hamilton 1977b). Lewontin shot back equally pugnaciously, arguing that Hamilton himself was responsible for his 'fair share of vulgar Darwinism' (Lewontin 1977b). Hamilton, Lewontin suggested, was no Darwin and Dawkins was no Huxley.

It also says something about the character of the two books that early on they attracted the attention not only of biologists, but also that of philosophers. Up until this point, the philosophy of science was very much centred around the philosophy of physics (see Mayr 1969 for an early version of this complaint). Anyone going through the defining texts of Karl Popper and Thomas Kuhn looking for insights into the nature of biological theories would be left wanting (although see Niemann 2014 for an account of Popper's 1986 Medawar Lecture on evolution). Today, the situation is very different. In an introductory essay accompanying the 50th anniversary edition of Kuhn's *The Structure of Scientific Revolutions*, the leading philosopher of science Ian Hacking describes biology as having replaced physics as science's 'top dog' (Hacking 2012, p. xv).

Both *Adaptation and Natural Selection* and *The Selfish Gene* were instrumental in the emergence of philosophy of biology as a distinct field of study. Elliott Sober, one of the scholars who defined the field, has described how he came to the philosophy of biology after being intrigued by the philosopher William Wimsatt's review of *Adaptation and Natural Selection* (Wimsatt 1970; Marshall 2016). Sober would later go on to grapple with many of the same issues in his influential *The Nature of Selection* (Sober 1984). Other field-defining books, such as Elisabeth Lloyd's *The Structure and Confirmation of Evolutionary Theory* (Lloyd 1988), Robert Brandon's *Adaptation and Environment* (Brandon 1990), and Daniel Dennett's *Darwin's Dangerous Idea* (Dennett 1995), all dedicated large chunks to issues that had been raised by Williams and Dawkins: causality, altruism and selfishness, and levels of selection.

Combined, these issues have spawned a rich and sprawling literature. Over the years, proponents and critics of the gene's-eye view have also developed a vocabulary that can bewilder. This chapter is dedicated to clarifying and untangling these terms.

To start, because the gene's-eye view envisions the history of life as a struggle between competing selfish genes, I will outline how this definition of a 'gene' differs from that used in molecular biology. Whereas in molecular biology, a gene has often been thought of as encoding a particular RNA or protein that then has a function, Williams and Dawkins advanced a definition a gene is defined as any part of a chromosome that is not broken up by recombination and is therefore passed on intact across generations. This definition is sometimes called the replicator concept, where replicators are entities whose structure is passed on intact across generations. Replicators are then complemented with vehicles, which are cohesive wholes that interact with the environment (a role typically played by individual organisms) and so cause replicators to be passed on. Thus, replicators cover one specific aspect of selection, transmission of hereditary information, whereas vehicles cover the other, ecological aspect.

Next, both Dawkins and Williams emphasized that the key property of a gene is not its physical properties, but its informational content. Defining genes in this non-material way was central to the development of the concept of memes, and I will include a brief discussion of this. Although the last few years have seen the term meme come to be associated with the rise of social media, Dawkins coined it in the last chapter of *The Selfish Gene* to serve as a unit of cultural transmission, a cultural replicator parallel to that of genes in organic evolution. I will show why despite the attempts to free replicators from the constraints imposed by the material basis of genes, those physical details are in fact key to understanding why the gene's-eye view has been so successful in the study of organic evolution but has failed to have a similar influence on the study of cultural evolution.

Finally, I will end the chapter by comparing how the concept of replicators and vehicles holds up next to other attempts to develop abstract formulations of evolution and natural selection, such as Lewontin's principles, and how the gene's-eye view fits into the contemporary major transitions research programme.

2.2 What is a selfish gene?

Under the gene's-eye view, fundamental terms are often defined in unusual ways. In a series of publications, David Haig has argued that accounting for these unusual definitions is key to understanding and getting the most out the concept (Haig 1997, 2006a, 2012, 2020). In Chapter 1, I discussed how Fisher introduced a radically expanded notion of the environment. Here, I will take a starting point in Haig's work to outline how another familiar term—the gene—took on a special meaning. Haig's telling of the memetic history of the gene begins not with Gregor Mendel, as deserving as he is of his place in the history of genetics, but with the Dane Wilhelm Johannsen. In 1910, Johannsen had been invited to attend the meeting of the American Society of Naturalists, which that year took place in Ithaca, New York. Despite being unable to attend in person, he nevertheless submitted a contribution to the society's journal and the paper appeared the following year (Johannsen 1911). His central motivation for writing the paper was that he thought that the study of heredity was hampered by an outdated terminology. In particular, he considered biometricians, such as Karl Pearson, to be too focused on the observable correlation between parents and offspring and to pay too little attention to the mechanisms that might cause the correlation. He contrasted this 'transmission conception' of heredity with his own 'genotype conception' and argued that shifting from the former to the latter was crucial in order to turn the study of heredity into an 'exact science'.

Building on the recent rediscovery of the work of Mendel, Johannsen introduced the distinction between the observable traits, which he called 'phenotype', and the heritable factors, which he dubbed 'genotype'. A gene, then, was: 'nothing but a very applicable little word, easily combined with others, and hence it may be useful as an expression for the "unit-factors," "elements" or "allelomorphs" in the gametes, demonstrated by modern Mendelian researches' (Johannsen 1911).

Since Johannsen, 'gene' has taken on a diversity of more precise meanings (Falk 1986; Griffiths and Stotz 2013; Kampourakis 2017). In his Nobel prize lecture delivered some 20 years after Johannsen had coined the terms phenotype and genotype, Thomas Hunt Morgan, winner of 1933 prize in Physiology or Medicine for his work on the role of chromosomes in heredity, noted that: 'There is no consensus of opinion amongst geneticists as to what the genes are—whether they are real or purely fictitious' (Morgan 1935).

The long-standing lack of a common definition of what a gene is has resulted in a number of situations of biologists talking past each other. The risk of mutually frustrating conversations seems to be particularly high when evolutionary and molecular biologists get together. One prominent example of such an incident is the molecular biologist Gunther Stent's reaction to reading Dawkins's definition of a gene in *The Selfish Gene*. Dawkins, following Williams in *Adaptation and Natural Selection* (Williams 1966, p. 24), defined a gene as: 'any portion of chromosomal material that potentially lasts for enough generations to serve as a unit of natural selection' (Dawkins 1976, p. 28). To Stent, the great sin of the Williams–Dawkins's gene concept was its vagueness on the molecular details. In his review of *The Selfish Gene,* he described the definition as 'denaturing the meaningful and well-established central concept of genetics into a fuzzy and heuristically useless notion' (Stent 1977).

In general, evolutionary biologists have typically showed little concern for the molecular intricacies of genes. After all, population genetics was developed long before the material basis of heredity was determined to be a nucleic acid and with only a meagre understanding of the relationship between genotype and phenotype. In so doing, they used a gene concept much closer to Mendel's original 'factors'. Here, a gene is simply something that is statistically associated with a difference in phenotype. Alfred Sturtevant, a student of Thomas Hunt Morgan and the first to construct a genetic map of a chromosome, summarized it like this:

> All that we mean when we speak of a gene for pink eyes is, a gene which differentiates a pink eyed fly from a normal one—not a gene which produces pink eyes per se, for the character pink eyes is dependent on the action of many other genes.   (Sturtevant 1915)

Sturtevant's comment has been conceptualized in various ways. Kitcher and Sterelny (1988), for example, referred to it as 'genes as difference makers'. The idea that a gene's effect is revealed only by comparing it with an alternative also features in Fisher's thinking. Recall from Chapter 1 that Fisher calculated an allele's effect by the partial regression of a given phenotype on the number of alleles present (i.e. 0, 1, or 2 in a diploid). To capture this aspect of genes, Moss (2003) introduced the gene concept Gene-P, the preformationist gene that predicts phenotypes. Gene-P is in contrast with the developmental Gene-D, which lies closer to the use of the term by molecular biologists, where the gene is a material thing defined by its molecular sequence. Whereas Gene-P would be familiar to Mendel and Johannsen, advances in the past half-century would render a paper on Gene-D incomprehensible to them. Lu and Bourrat (2018) use many similar arguments in discussing the relevance of recent empirical discoveries of epigenetic inheritance for the definition of genes but suggest the terms evolutionary genes (following Griffiths and Neumann-Held 1999) and molecular genes. The key take-home message is that the gene's-eye view wants to talk about genes in an abstract way and happily accepts a bit of fuzziness regarding their physical basis.

The gene's-eye view's way of conceptualizing genes also has implications for how we think about phenotypes. Instead of belonging to organisms, the traditional view of most biologists, phenotypes belong to genes. 'A gene's effects are its phenotype', as Dawkins (1982a, p. 4) put it. A gene's effect can therefore only be understood in comparison to some alternative allele (Haig 2012; Lu and Bourrat 2018). If there is no alternative, there is, by definition, no phenotype. While this may initially seem strange, it comes close to Johannsen's original definition of a phenotype, which explicitly considered distinguishable 'types' of organisms.

What happens to the notion of 'genes as difference makers' when genes lack phenotypic effects? This is not just of philosophical interest but becomes a real issue when genome-wide association studies reveal little to no signal for most parts of the genome (Noble and Hunter 2020). One response would be that a lack of effect implies that such genes are not genes at all. Alternatively, one can take it to mean that those genes will just not be subject to selection and that their fate in the population will be determined by genetic drift.

Finally, adopting the gene's-eye view's gene concept brings us back to Fisher's expanded notion of the environment. Whereas molecular biologists and ecologists are united in defining the environment as that beyond the physical boundaries of the individual organism, from an gene's-eye view it includes that plus the other alleles at the same locus, the rest of the genome, and the gene pool in the population: in essence, 'all parts of the world that is shared by the alternatives being compared' (Haig 2012; see also Sterelny and Kitchner 1988).

One benefit of transferring ownership of the phenotype from the individual organism to the gene is that it allows one to consider phenotypes beyond the body of the organism, such as extended phenotypes (see Chapter 5).

2.2.1 How long is a selfish gene?

A question that arose early about the gene's-eye view gene was how long does a 'portion of chromosomal material' need to be to count as a gene? Some, such as Peter Godfrey-Smith consider the lack of clear lengths to be a fatal flaw of the gene's-eye view gene concept (Godfrey-Smith 2009, pp. 135–139). Again, Dawkins and Williams appear rather relaxed about this point. Dawkins repeatedly emphasized that a gene can be of arbitrary length (Dawkins 1976, p. 35, 1982a, p. 87) and Williams remarked that:

> Various kinds of suppression of recombination may cause a major chromosomal segment or even a whole chromosome to be transmitted entire for many generations in certain lines of descent. In such cases the segment or chromosome behaves in a way that approximates the population genetics of a single gene. (Williams 1966, p. 24)

By this definition, then, the non-recombining parts of the Y and W chromosomes, as well as the entire genomes of mitochondria and asexually reproducing organisms can be considered one gene (Haig 2012). A sexual genome, however, is not a gene according to the gene's-eye view, for recombination and crossing-over during meiosis breaks it up into multiple fragments.

The answer to the question of the length of a selfish gene largely hinges on the extent of linkage disequilibrium. In a field notorious for its convoluted vocabulary, linkage disequilibrium is one of the worst offenders. In short, it simply means that the frequency of association between alleles at different loci is different from random. Population geneticists like Montgomery Slatkin argued that the presence of widespread linkage disequilibrium meant that it made little sense to talk about genes in this way, that is as individual units during sexual reproduction (Slatkin 1972).

The gene of the gene's-eye view has also been challenged from the opposite direction. If there is extensive recombination, no chromosomal portion will be transmitted intact for long enough to act as a unit of selection. At the limit, this would mean that there are exactly four genes, the four constituent bases of nucleic acids: adenine, thymine, guanine, and cytosine (Griffiths and Sterelny 1999, p. 80). Dawkins refers to this objection as the 'selfish nucleotide theory' (Dawkins 1982a, pp. 91–92). His counterargument centres on emphasizing genes as 'difference makers'. Because nucleotides cannot be said to have a phenotype, in the way a longer DNA segment can, Dawkins argues it still makes sense to talk about genes:

> The single nucleotide (…) cannot be said to have a phenotypic effect except in the context of the other nucleotides that surround it in its cistron [the stretch of DNA that encodes a single polypeptide]. It is meaningless to speak of the phenotypic effect of adenine. But it is entirely sensible to speak of the phenotypic effect of substituting adenine for cytosine at a named locus within a named cistron. (…) Unlike a nucleotide, a cistron is large enough to have a consistent phenotypic effect, relatively, though not completely, independently of where it lies on the chromosome.   (Dawkins 1982a, pp. 91–92)

Another way to respond to the 'selfish nucleotide' charge is to concede and embrace it. In *Darwinian Reductionism: Or, How to Stop Worrying and Love Molecular Biology*, Alex Rosenberg attacks what he regards as an 'untenable dualism' characterizing much contemporary philosophy of biology (Rosenberg 2006). To Rosenberg, the dualism in question arises because most philosophers of biology, and evolutionary biologist for that matter, are physicalists. That is, they believe everything in the universe—matter and mind—is made up of physical things. At the same time, however, they are often anti-reductionist in the sense that they seek explanations at the level of biological phenomena—genes, organisms, populations—rather than at the level of chemistry and physics. Indeed, the inability to reduce biology to special cases of physical laws is often presented as key to the autonomy of biology as its own scientific disciple (e.g. Mayr 2004, pp. 26–28). Such a dualism, Rosenberg argues, opens up the door to ideas like intelligent design and vitalism (the idea that living organism are made up of some non-physical matter making them fundamentally different from non-living things).

To get around this, Rosenberg presents a spirited defence of reductionism. Of particular relevance here is his argument that natural selection should be considered a physical law that can operate at one or more lower levels of aggregation. At the most basic level, selection will act on aggregates of atoms and macromolecules and will favour those aggregates that are the most stable and fastest at replicating. In principle, then, the same process will apply all the way up to cells, individuals, and groups. Rosenberg's reasoning shows that the 'selfish nucleotide' is not necessarily as absurd as it might appear at first glance, but his argument also part of a larger, and for scientists slightly intimidating, discussion about reductionism in science in general and biology in particular. See, for example, Nicholson and Dupré (2018) for a very different perspective.

2.2.2 Is a selfish gene a token or a type?

The tension between treating genes as faithfully transmitted difference makers, agnostic about their material basis, or as physical object like a stretch of DNA with discrete boundaries, cuts across many debates about the gene's-eye view. One way to make sense of this disagreement is to use the philosophical distinction between type and token (Haig 2012, 2020; but see Mitchell 2003, pp. 63–74). To appreciate the difference between types and tokens, consider the two ways one can answer the question 'how many mitochondrial genomes does a human somatic cell contain?' The first answer would be 1, as the mitochondrial genome is strictly maternally inherited and therefore all mitochondrial genomes in the human body are genetically identical, bar any *de novo* mutations. The genome type thus occurs exactly once. The second approach to answering the question would be to count the number of mitochondria in the cell in question. The number of times the genome token appears will depend on the exact cell type, ranging from 0 in a red blood cell to around 2,000 in a liver cell.

Whether the gene's-eye view is best formalized using the type or token approach has been subject to debate. Gardner and Welch (2011) opted for the latter approach and used the mathematics of optimization theory to develop a model of the gene as an inclusive fitness-maximizing agent. Here, the gene is a physical object, a stretch of DNA. The central conclusion of this analysis is that the gene does not necessarily behave selfishly; it may also behave altruistically or spitefully towards other genes. They therefore argue that the only way to retain the 'selfish' in 'selfish genes' is to define selfishness in the trivial sense of being evolutionary successful (see also Noble 2011; Gintis 2014, 2016, pp. 185–224).

Haig has conceived of genes as types in his 'strategic gene' framework (Haig 2012, 2020). This approach takes seriously Dawkins's comment in *The Selfish Gene*: 'What is the selfish gene? It is not just one single physical bit of DNA (...) it is all replicas of a particular bit of DNA distributed throughout the world' (Dawkins 1976, p. 95). One consequence of the type approach is that it does not make sense to think of a gene as being physically located in an individual body. Instead, to Haig, the strategic gene is the collective of all tokens whose chance of transmission is affected by a given token. The actor token and the recipient token(s) may be physically in the same organism, but the approach works equally well if they are not (see discussion of greenbeard genes in Chapter 5).

Whether one prefers types or tokens is partly a matter of taste. The token approach has the benefit of a rich tradition of mathematical modelling of social evolution at the organismal level (Gardner 2014a). Thinking in terms of gene tokens having different inclusive fitness agendas may be particularly useful when studying genomic conflicts (see Gardner and Úbeda 2017). The type framework, in contrast, is closer to Dawkins's original argument. The notion that genes' ultimate goal is to increase their frequency in the population also only really makes sense if the genes are types (Okasha 2019), as only gene types (that is, alleles) can have a frequency. Again, this emphasizes the point made in Chapter 1, that a more accurate name for Williams and Dawkins's argument could be said to be the allele's-eye view.

In general, both Gardner's and Haig's approaches are in line with the general agential approach to evolutionary theory favoured by Dawkins (Okasha 2018). I will return to the agency concept, and its contentious role in the history of biology, in Chapter 3. With this discussion of genes in place, I now turn to the major conceptual advance enabled by taking the gene's-eye view: the distinction between replicators and vehicles.

2.3 Replicators and vehicles

> Like [the pretentious social climber in Molière's 1670 play *The Middle Class Aristocrat*] Monsieur Jourdain, who was astonished to discover that he had been speaking prose all his life, Dawkins may well be surprised to discover that he had committed an act of metaphysics.   (Hull 1981)

To the philosopher David Hull, Dawkins's act of metaphysics was the introduction of the replicator and the associated argument that evolution by natural selections involves two central entities: replicators and vehicles. Hull was the first to clearly state that the evolutionary process could be conceptualized in this way (though he used the term interactor, rather than vehicle; Hull 1980, 1981). Williams had earlier distinguished between 'genic selection and organic adaptation' (Williams 1966, p. 124) and Hull's insight is also implicit in a much-quoted passage of *The Selfish Gene* about how genes control bodies:

What was to be the fate of the ancient replicators? They did not die out, for they are past masters of the survival arts. But do not look for them floating loose in the sea; they gave up that cavalier freedom long ago. Now they swarm in huge colonies, safe inside gigantic lumbering robots, sealed off from the outside world, communicating with it by tortuous indirect routes, manipulating it by remote control. They are in you and in me; they created us, body and mind; and their preservation is the ultimate rationale for our existence. They have come a long way, those replicators. Now they go by the name of genes, and we are their survival machines.   (Dawkins 1976, p. 25).

This paragraph is notoriously purple, even after Michael Rodgers toned down its first draft (Dawkins 2013a, p. 277). Personally, I have always preferred the limerick version, first introduced by Dawkins at a conference banquet and retold in *The Ancestor's Tale*:

> An itinerant selfish gene
> said 'bodies a-plenty I've seen.
> You think you're so clever,
> But I'll live forever.
> You're just a survival machine.   (Dawkins 2004a, p. 61)

The central message in both passages is that replicators and vehicles play separate roles in the evolutionary process. Replicators are any entities whose structure and information are copied and faithfully transmitted from parent to offspring, forming lineages across generations. A successful replicator has three properties: longevity, fecundity, and copy-fidelity (Dawkins 1978). In organic evolution, the role of replicator is usually played by genes.

Vehicles, on the other hand, what Dawkins referred to above as survival machines, are the entities in which genes are bundled together and that interact directly with the external environment. This role is typically filled by individual organisms but may also be carried out by cells and, more rarely, by groups. Under this formulation, then, natural selection is a process in which some vehicles are more successful than others, leading to the survival and proliferation of their replicators. To Hull and Dawkins, replicator survival and vehicle/interactor selection are two sides of the same coin.

Not everyone has seen merit in the replicator–vehicle distinction. Michael Ghiselin called it a 'misleading metaphor in support of a dubious metaphysical thesis' (Ghiselin 1997, p. 147) and Conor Cunningham dismissed it as 'quasi-Cartesianism' (Cunningham 2010, p. 65; see also Goodwin 1994, pp. 29–34 for a similarly flavoured complaint). In an early critique of the gene's-eye, Stephen Jay Gould stated that selection cannot 'see' individual genes, only individual organisms (Gould 1977; see also Mayr 1963, p. 184 for a related argument against population genetics). That is, selection does not 'care' why an individual is successful, just that it is. To use Robert Brandon's terminology, the phenotype 'screens off' the underlying genotype, making only the former visible to selection (Brandon 1990, pp. 83–85).

Thinking in terms of replicators and vehicles shows why this critique is mistaken for at least two reasons. First, selection can clearly see individual genes in the case of selfish genetic elements and other forms of genomic conflict (see Chapter 5). Second, Dawkins repeatedly emphasized that replicators need vehicles to be transmitted: '[replicators do not] literally face the cutting edge of natural selection. It is their phenotypic effects that are the proximal subjects of selection' (Dawkins 1982b). It is fair to say, however, that Dawkins certainly favoured replicators, and his disagreement with Gould was partly a consequence of Gould's interest in vehicles over replicators (Istvan 2013). When reading *The Selfish Gene* one can also easily come away with the impression that only replicators matter. Though *The Extended Phenotype* is more balanced, the stated purpose of that book is to undermine the idea of organisms as a useful concept in evolutionary biology. As Dawkins would later note that he created the vehicle concept 'not to praise it, but to bury it' (Dawkins 1994b). His preference for replicators is also reflected in the passivity inherent in the term vehicle. A vehicle, Dawkins argued, 'can be regarded as a machine programmed to preserve and propagate the replicators that ride inside it (Dawkins 1982a, p. 295)', and its passivity is 'paradoxically why vehicle is a better name than Hull's "interactor" (Dawkins 1994). Interactor comes too close to the (messy) truth and therefore does not merit a helpfully decisive burial' (Dawkins 1994b). In contrast, Hull preferred interactor as it was more balanced in its relation to replicators (Hull 1980, 1981). Rather than being mere 'survival machines' and 'lumbering robots' for replicators, interactors play an active role in their environment. Vehicles and interactors, though related, are thus not equivalent.

2.3.1 Lloyd's four questions and the immortality of replicators

Recall from Chapter 1 that proponents of the gene's-eye view identify the appearance of design in the living world as the central problem of evolutionary biology. To Dawkins, the way to address this problem is to ask 'when we say that an adaptation is "for the good of" something, what is that something?' (Dawkins 1982b). That is *cui bono?* Or what is the beneficiary of natural selection? This is what Elisabeth Lloyd referred to as a 'specific ontological issue of benefit' or the beneficiary question (Lloyd 2017). As a philosopher, Lloyd did an enormous job in clearing up and sharpening the debate about units and levels of selection, a debate that was long plagued by strong disagreements and participants talking past each other (Lloyd 1988, 1989, 1992, 2017).

Lloyd's crucial contribution was to distinguish the question of benefit from three other questions that can be asked about the units and levels of selection. In any given situation, Lloyd suggests that four different questions highlight separate aspects of the selective process:

1. What is the replicator?
2. What is the interactor?
3. Where is adaptation manifested?
4. What is the beneficiary of selection?

The first two questions correspond to the replicator and vehicle/interactor distinction discussed above. Another way of asking the third one would be: 'When a population evolves by natural selection, what, if anything, is the entity that does the adapting?' (Sober 1984, p. 204). This seems straightforward enough. For example, one may argue that selection is really happening at the replicator level, but that it results in adaptation at the individual or even group level. Alternatively, one may require that adaptation at a given level is the product of selection at that level (Sober and Wilson 2011).

The gene's-eye view is based on the conviction that the beneficiary question is the most important one. With this in mind, it is easier to see why Williams and Dawkins were happy to accept such fuzziness around their gene definition. The gene's-eye is primarily interested in the evolutionary origin of complex adaptations and trades in hard-won molecular details of the structure of genes, for the ability to work out the logic of natural selection. This trade-off pays off when natural selection results in phenomena that makes little sense from the perspective of individual organisms, such as worker sterility in eusocial insects or genomic conflicts. When those molecular details matter, such as in many questions of developmental biology, the price of the trade-off may be too high.

The centrality that they placed on the beneficiary question also helps make sense of why Dawkins and Williams put so much emphasis on the unique properties of replicators. Chief among these properties is that replicators are potentially immortal, as opposed to the transient nature of vehicles. When Dawkins was looking for a publisher for *The Selfish Gene,* he met with the influential editor Tom Maschler, then at Jonathan Cape. Maschler liked the manuscript but not the title and suggested The Immortal Gene as an alternative. This change would capture much of the same message of the book but avoid the word 'selfish', which Maschler considered a 'down word'. Dawkins would later think Mashler might have been right (Dawkins 2006a, p. vii).

Replicators are the beneficiaries of natural selection because their immortality means that they are the entities that survive the evolutionary process. Both Williams (1966, pp. 23–24) and Dawkins (1976, p. 34) argued that while individual organisms measure their life span in decades, replicators do so in thousands and millions of years. You cannot have evolution within a single generation and only replicators survive across generations. The replicators that can be said to form lineages across generations are those in sex cells, what Dawkins called 'active germ-line replicators' (Dawkins 1982a, p. 83). Any mutations in somatic cells are considered dead-end replicators. Active germ-line replicators are repeatedly subject to natural selection, generation after generation. Organisms, however, are not. Whereas a mutation in a successful replicator is transmitted, a change in an organism's phenotype—a plant having one of its leaves chewed off by a caterpillar, for example—is not. As a consequence, only replicators have the evolutionary persistence to be responsible for the cumulative selection required for evolution to result in adaptations.

The argument that only replicators, not organisms, form lineages can be traced back to August Weismann's idea of the 'continuity of the germ-plasm' (Haig 2007). It was Weismann who introduced the distinction between immortal germ line cells that produce sperm and eggs and mortal somatic cells that die with the organism (Weismann 1892). Weismann's distinction was crucial to the rejection of the Lamarckian idea of the inheritance of acquired characteristics, as only changes to germ cells will be transmitted to the next generation. In retrospect, his separation between the mortal body and the immortal germ-line can be also seen as an early version of the replicator–vehicle distinction. In this vein, Dawkins clearly acknowledges his intellectual debt to Weismann early in *The Selfish Gene* (Dawkins 1976, p. 11) and has later called him the 'father of the selfish gene concept' (Dawkins 1994a).

Today, the emphasis on active germ-line replicators appears overly restrictive. There is increasing evidence that parents faithfully pass on more than their genes (Bonduriansky and Day 2018). Moreover, while only in its infancy around the publication of *The Selfish Gene*, the idea that cancer is best thought of as an evolutionary phenomenon is now clearly recognized (Stearns and Medzhitov 2015, chapter 6; Aktipis 2020). This is true even if it only involves somatic cells evolving within one organismal generation.

Genes are also not the only biological entity that have lifespans in the millions of years, so do two others: species and traits. A crucial difference between species and replicators is that replicators affect what vehicles survive and reproduce and therefore can be said to have agency in the evolutionary process. Species cannot be said to play such a causal role (Hampe and Morgan 1988; Lloyd 2017). Instead, they passively benefit from the process of natural selection.

The trait challenge rejects the premise that cumulative selection requires something to form lineages. By this argument, selection only requires that something is persistently present across multiple generation, a requirement satisfied by many phenotypic traits. Take, for example, a plant trait like the density of trichomes that prevents a leaf from being eaten by caterpillars. The trait 'trichomes' may be present in a population, though manifested on different individuals, for multiple generations, and so satisfies the persistence requirement as outlined by Williams and Dawkins. Indeed, studying traits across phylogenetic trees is at the heart of the comparative method in evolutionary biology (Harvey and Pagel 1991). Evolutionary longevity is thus not unique to replicators.

In response, proponents of the gene's-eye view have therefore emphasized another aspect that distinguishes genes from other biological entities: information. Replicators are the only entities that survive in the evolutionary process because they should be thought of not as physical objects, but as units of information. As Dawkins put it in *The Blind Watchmaker*:

> It is raining DNA outside. On the bank of the Oxford canal at the bottom of my garden is a large willow tree, and it is pumping downy seeds into the air. (…) [spreading] DNA whose coded characters spell out specific instructions for building willow trees that will shed a new generation of downy seeds. (…) It is raining instructions out there; it's raining programs; it's raining tree-growing, fluff-spreading, algorithms. That is not a metaphor, it is the plain truth. It couldn't be any plainer if it were raining floppy discs.   (Dawkins 1986a, p. 111)

This again shows why the type, rather than token, concept of genes makes more sense for the gene's-eye view. A gene token is no more immortal than an individual organism.

There is an extensive literature on the concept of genes as information (see e.g. Maynard Smith's 2000 paper in *Philosophy of Science* and comments from Godfrey-Smith 2000a, Sarkar 2000, Sterelny 2000, and Winnie 2000). Williams argued that 'evolutionary biologists have failed to realize that they work with two more or less incommensurable domains: that of information and that of matter'. This message was a major theme of his 1992 book *Natural Selection: Domains, Levels, and Challenges.* Towards the end of his career, Williams would even say that he considered his work on the informational aspects of the gene to be his major contribution to evolutionary biology; if it was not for him, someone else would have come up with the gene's-eye view (Williams 1996b, p. 45). Similarly, a major reason for why Dawkins preferred the term replicator over gene was because it disentangled the concept from any material basis and allowed for evolution by natural selection of other kinds of replicators (Dawkins 1982b; Dawkins 2006b, p. 228). In particular, Dawkins was interested in the possibility of cultural replicators, what he dubbed memes.

2.4 Memes

Attempts to apply Darwin's theory to cultural matters began soon after the publication of the *Origin of Species* (reviewed in Lewens 2015 and Lewens 2018). In 1880, the American philosopher and pioneering psychology educator Williams James wrote:

> A remarkable parallel, which to my mind has never been noticed, obtains between the facts of social evolution and the mental growth of the race, on the one hand, and of zoological evolution, as expounded by Mr Darwin, on the other.   (James 1880, p. 441; quoted in Lewens 2018)

James was writing in the context of a debate about 'great men' in history spurred by Herbert Spencer. Since then, many others have tried to develop more general accounts of cultural evolution. Early modern attempts came from population geneticists like Cavalli-Sforza and Feldman (1981), anthropologists like Boyd, and biologists like

Richerson (Boyd and Richerson 1985). Though not the first or most influential, memetics may well be the most famous attempt to use insight from evolutionary theory to study cultural change (Aunger 2001; Lewens 2018).

Whereas the rise of social media has led to the term meme becoming associated with both funny cats and right-wing frogs, its origin lies elsewhere. The term was first defined in the last chapter of the first edition of *The Selfish Gene*. (Though, unbeknownst to Dawkins, there were some earlier uses of the term; see Laurent 1999.) Dawkins chose the word meme because it resembled the word gene:

> We need a name for the new replicator, a noun that conveys the idea of a unit of cultural transmission, or a unit of imitation. 'Mimeme' comes from a suitable Greek root, but I want a monosyllable that sounds a bit like 'gene'.   (Dawkins 1976, p. 192)

The central claim of memetics is that because organic evolution requires something to play the role of replicator, so does a theory of cultural evolution. Examples of memes suggested by Dawkins included 'tunes, ideas, catch-phrases, clothes fashions, ways of making pots or of building arches' (Dawkins 1976, p. 206). The idea is that memes will spread from mind to mind, with contagious ideas becoming more common. The success of a given meme—say a new scientific or religious idea about how the world works—will depend on how well it fits the cultural environment in which it arises. In this way, the success of a meme parallels that of a gene. While the gene's-eye view likes to think of individual genes, the only way to do this is by taking into account what other genes are in the population. Memes can also spread together in so-called memeplexes; examples include combinations of ideas, such as a political ideology.

Just as the gene's-eye view opens our eyes to the idea that not all genes may be working for the same purpose, thinking of culture in terms of memes helps us see how harmful ideas may spread. Not all memes will be beneficial to the vehicle that houses them. In the same way that selfish genetic elements can spread despite the fitness cost to the individual organisms, so can harmful memes. This is at the heart of Dawkins's description of religion as a virus of the mind (Dawkins 1993b; Dawkins 2006b, pp. 218–219).

Memes as a theory of cultural evolution attracted both supporters and critics. Daniel Dennett is a supporter and used it to develop a theory of mind in *Consciousness Explained* (Dennett 1991). So is physicist David Deutsch, who incorporated replicators (both genes and memes) as one of the four strands of his 'theory of everything' in *The Fabric of Reality* (Deutsch 1997). The most ambitious attempt to develop a comprehensive theory of memes remains Susan Blackmore's *The Meme Machine* (Blackmore 1999), a book with an almost evangelical tone that came with a foreword by Dawkins. More recently, the case for memetics was laid out by Stewart-Williams (2018). A journal dedicated to the study of memes, *The Journal of Memetics—Evolutionary Models of Information Transmission,* briefly existed online from 1997 to 2005.

Critics of memetics have often been harsh. Ernst Mayr, for example, dismissed it as little more than a renaming of the word 'concept' (Mayr 1997). The philosopher and historian of ideas John Gray, who has often incorporated insight from evolutionary theory in his work, went further, comparing it to intelligent design and describing it as 'a classic example of the nonsense that is spawned when Darwinian thinking is applied outside its proper sphere. (…) Talk of memes is just the latest in a succession of ill-judged Darwinian metaphors' (Gray 2008).

The idea of memes is an attempt to free the replicator concept from the specific physical features of genes, yet those details show why the replicator concept has been so influential in the study of organic but not cultural evolution (Claidière and André 2012; Lewens 2015). The first problem concerns where the boundaries for what is considered a single meme should be drawn (Daly 1982). This question arises when discussing genes as well (see 2.2. What is a selfish gene?), but it can be resolved by considering clear biological processes like recombination and crossing over: a biological replicator is that which is stably inherited across generations. This kind of stability is much harder to achieve with cultural entities (Sterelny 2006). Furthermore, the integrity of a gene can be disrupted by mutation. In organic evolution, mutations are essential for introducing new variation for selection to act upon. A mutation rate that is too high, however, drowns selection with random variation. Memes are therefore much likelier to fail to satisfy the trio of properties of successful replicators: longevity, fecundity, and copy-fidelity.

Next, memes, as opposed to genes, do not form lineages. Cultural ideas reproduce in the sense that they are created again and again, and thus there exists a meaningful way in which it is possible to talk about a causal link between different creations, but not the sense that one is a copy of another, like a gene (Wimsatt 1999; Sperber 2000). Whereas a given gene token can in principle be traced back to the very point where the mutation first arose, memes rarely or never work like this.

A final issue is how competition between memes is supposed to work. Dawkins anticipated this problem: 'memes seem to have nothing equivalent to chromosomes and nothing equivalent to alleles' (Dawkins 1976, p. 211). Competition among alleles at the same chromosomal location is how the gene's-eye envisions natural selection. Mendelian inheritance, with occasional deviations for selfish genetic elements, is thus crucial to the gene's-eye view in the organic arena. In culture, what plays that role? One response to this criticism is to point out that in the early history of life, biological replicators were not organized along chromosomes but may have floated free in protocells. Whereas meiosis is integral to how inheritance works in contemporary sexually reproducing organisms, it is in no way essential to a theory of evolution, not even one based on the replicator concept. Imposing this particular requirement on memetics is therefore unfair.

The contemporary study of cultural evolution is a thriving field (Lewens 2015), though not without critics (Ingold 2007). Memetics played an important role raising the field's profile and it especially highlighted that cultural practices may persist not because of the benefit to the individual or the group performing them, but because they have qualities that make them good at persisting regardless of consequences. Yet, the above shortcomings mean that other theoretical frameworks now dominate the field.

2.5 General formulations of evolution by natural selection

In a 2015 essay, Dawkins argued that:

> ...if there is life elsewhere in the universe, it will be Darwinian life. (...) This all comes about because at some point in history, about 4 billion years ago, a replicating entity arose, not a gene as we would now see it, but something functionally equivalent to a gene, which because it had the power to replicate and the power to influence its own probability of replicating, and replicated with slight errors, gave rise to the whole of life. (...) So for me, the replicator, the gene, DNA, is absolutely key to the whole process of Darwinian natural selection. So when you ask the question, what about group selection, what about higher levels of selection, what about different levels of selection, everything comes down to gene selection. Gene selection is fundamentally what is really going on.   (Dawkins 2015b, pp. 1–2)

In so doing, he touched upon two issues that have played important roles in disagreements over the gene's-eye view. The first is the assertion that evolution requires something to play the role of replicator. Above and elsewhere Dawkins has argued that replicators will be a feature of life, wherever it may be found in the universe (Dawkins 2015a, p. 331). The Dawkins–Hull formulation of replicators and vehicles fit into a larger, more general, discussion about how best to articulate the principles of evolution by natural selection in the abstract. In contrast to Dawkins and Hull, many authors have considered replicators to be a limited case and have tried to articulate broader principles. The second is whether gene selection a process or a perspective. Above, Dawkins gives the impression that selection at the gene level is a more factually true way to describe evolution than selection at the organismal or group level. This idea goes back all the way to the earliest writings of Dawkins, but he has also on occasion argued in favour of the weaker claim that the gene's-eye view is but one way of viewing the facts of evolution, no more true than an organism-centred perspective.

Below, I tackle the issue of abstract formulations of evolution by natural selection and I will return to the process versus perspective issue in Chapter 3.

2.5.1 Lewontin's Principles and limits of the replicator–vehicle approach

Darwin originally formulated evolution by natural selection as individual organisms engaging in a 'struggle for life':

> Owing to this struggle for life, any variation, however slight and from whatever cause proceeding, if it be in any degree profitable to an individual of any species, in its infinitely complex relations to other organic beings and to external nature, will tend to the preservation of that individual, and will generally be inherited by its offspring. (…) I have called this principle, by which each slight variation, if useful, is preserved, by the term Natural Selection.   (Darwin 1859, p. 61).

It was quickly realized, however, that the abstract nature of the principles of selection meant that it could operate at multiple levels and even within organisms (see e.g. Huxley 1878). The classic articulation of this point was made by Lewontin (1970) and is now sometimes referred to as Lewontin's Principles (Brandon 2019). Lewontin stated that evolution by natural selection requires three things:

1. Phenotypic variation. For a population to evolve, individuals must vary in phenotype, such as morphology, physiology, or behaviour.

2. Phenotypes must differ in fitness. Variation in phenotypic traits must be associated with a difference in ability to survive and reproduce.

3. Fitness is heritable. The fitness of parents must be causally correlated with that of their offspring.

John Maynard Smith developed an account very similar to Lewontin's and argued that the evolutionary process was built on a triptych of 'multiplication, variation and heredity' (Maynard Smith 1987). Both Lewontin's and Maynard Smith's accounts function as conditional statements. If a population satisfies these conditions, evolution by natural selection may occur. This approach has recently been expanded to the concept of Darwinian Populations by Peter Godfrey-Smith (Godfrey-Smith 2009; see also reviews by Pigliucci 2009; Dennett 2011; Queller 2011; Sterelny 2011 and reply by Godfrey-Smith 2011 for an overview of the debate to which this concept has led).

Godfrey-Smith classified his and Lewontin's recipe approach as the 'classic' tradition to defining natural selection (Godfrey-Smith 2009, p. 4). Together with life's hierarchical organization (genes in genomes, genomes in cells, cells in organisms, organisms in social groups), this classic tradition is what gave rise to the levels-of-selection debate to which the gene's-eye view owes part of its existence. And it is in contrast to this traditional view that the replicator–vehicle approach of Dawkins and Hull should be understood. Their approach requires two entities, replicators and vehicles, whereas the recipe version requires only one. This difference in and of itself arguably means that the recipe formulation is more general (Okasha 2008b). That being staid, it does not take away from the fact that the replicator–vehicle distinction has proved very useful for understanding a wide variety of evolutionary questions. Instead, the replicator concept has other, more serious, weaknesses.

One such weakness is that the recipe approach shows that genes are 'optional' to the evolutionary process. Whereas evolution requires some sort of heritability, that is enough parent–offspring correlation in traits to lead to cumulative evolutionary change, the mechanism need not be the particular kind of inheritance found on this planet (Godfrey-Smith 2000b). Okasha (2006, p. 15) referred to this as 'Gould's paradox', following Gould's argument that because Darwin did not have a functioning theory of inheritance, but clearly understood evolution, it does not make sense to treat replicators as fundamental (Gould 2002, p. 613). Again, however, the fact that evolution would work even under blending inheritance (though it would require an extremely high mutation rate to avoid running out of variation; Bulmer 2004) does not mean that the replicator–vehicle concept is not helpful.

Take, for example, the concept of kin selection. Some authors have argued that the reason why Darwin himself did not develop a theory of kin selection was that he was operating in a pre-Mendelian intellectual world (e.g. Borello 2010). The main evidence for this claim is Hamilton's reliance on a gene's-eye view when introducing the idea (Hamilton 1963). Yet, as Gardner (2011) has convincingly shown, this need not be the case. By deriving a kin selection model under the assumption of blending inheritance, Gardner demonstrated that the logic of kin selection does not rely on particulate inheritance. The key insight of his analysis is that the idea of kin selection could in principle have been developed even in the absence of an understanding of Mendelian genetics. Gardner's conclusion comes with the caveat that the analysis becomes more 'tortuous' under the assumption of blending, rather than particulate, inheritance. Thus, just as for evolution in general, inheritance through genes is not required to understand kin selection, but it certainly seems to make it easier to make sense of many real life examples.

Another weakness is that the properties of replicators and vehicles are themselves a product of selection. A requirement under the Dawkins–Hull framework is that some entities have high enough copy-fidelity to play the role of a faithfully transmitted replicator. Some critics of the gene's-eye view (e.g. Ball 2018) think that the mutation rate in the earliest genes was too high for them to play the role of a replicator. This objection is related to the issue identified already by Manfred Eigen (independent of debates over replicators and vehicles) that the mutation rate, measured in copying mistakes per base pairs, becomes a lethal issue before a self-replicating molecule can reach the length necessary to encode an enzyme that can correct such mutations (Eigen 1971). While many solutions have been proposed, the issue is hard to get around (Szathmáry 2006). Furthermore, the earliest vehicles, probably some sort of simple unicellular organism or proto-cell, were far from the cohesive entities implied by the vehicle concept (Buss 1987; Michod 1999). Thus, the replicator–vehicle framework suffers from the problem that its conceptual structure assumes some of the empirical observations (such as the fidelity of replicator transmission or cohesiveness of individual organisms) that it is supposed to explain (Griesemer 2000). As Denis Noble has emphasized, DNA can typically not do much on its own but requires the infrastructure of the cell to achieve anything (Noble 2018; see also Griesemer 2006). By this argument, the separation of replicators as active and vehicles as passive therefore gets things completely backwards.

The debate over units and levels of selection has since been transformed from taking the hierarchy of life for granted to being concerned with the evolutionary origins of this hierarchy. As Leo Buss argued, 'individuality is a derived character' (Buss 1987, p. 14). The fact that cells work together in multicellular organisms is not just a coincidence, but the product of a series of evolutionary events. Explaining the emergence of new levels in the hierarchy is the goal of the major transitions research programme (Maynard Smith and Szathmáry 1995; Michod 1999; Bourke 2011; West et al. 2015). How does the genes'-eye view fare in this new world of major transitions? To that I turn next.

2.5.2 The major transitions and the levels of selection debate

The major transitions research programme changed and re-energized the levels of selection debate. Instead of taking the hierarchical organization of life for granted, much effort is now devoted to explain its evolutionary origins (see Okasha 2006 and Bourrat 2021). The term 'major transitions' originally comes from John Maynard Smith and Eörs Szathmáry, who published *The Major Transitions in Evolution* in 1995 (and in 1999 a popular science version, or as Maynard Smith put it: 'a birdwatchers' version'; Maynard Smith and Szathmáry

1999; Piel 2019). The definition of the term has changed slightly over the years. Maynard Smith and Szathmáry (1995, pp. 3–4) defined them as transitions 'in the way in which genetic information is transmitted between generations' such that 'entities that were capable of independent replication before the transition can replicate only as part of a larger whole after it' (Maynard Smith and Szathmáry 1995, pp. 3–4). This definition resulted in eight transitions:

1. Replicating molecules → Populations of molecules in compartments

2. Independent replicators → Chromosomes

3. RNA as gene and enzyme → DNA + protein (the genetic code)

4. Prokaryotes → Eukaryotes

5. Asexual clones → Sexual populations

6. Protists → Animals, plants, fungi (cell differentiation)

7. Solitary individuals → Colonies (non-reproductive castes)

8. Primate societies → Human societies (language)

To Maynard Smith and Szathmáry, what unified these major transitions was that they changed how the evolutionary process itself operates by altering how heritable information is stored and transmitted. Not all of their major transitions, however, involved the coming together of previously independently replicating entities, now known as evolutionary transitions in individuality and what most contemporary researchers are concerned with (Clarke 2014; Herron 2021; but see McShea and Simpson 2011). The evolution of the genetic code is a good example of a change in the language, storage, and transmission of information that is not an evolutionary transition in individuality.

Major transitions in individuality share two central features (Maynard Smith and Szathmary 1995; Bourke 2011; West et al. 2015). First, transitions in individuality involve the emergence of cooperation among independent entities and lead to the formation of a new higher-level individual entity. Second, crucial to the functioning of this new level of individuality is the evolution of mechanisms to suppress conflict among lower-level entities.

Before *The Major Transitions in Evolution* was published, similar ideas had been in the air for some time. In 1974, John Tyler Bonner grouped the origin of the eukaryotic cell, multicellular organism, and social groups as examples of complexity being the product of independent units coming together to form new wholes (Bonner 1974). Next, Buss's *The Evolution of Individuality* was the first book-length treatment of the modern hierarchical view and it focused on the evolution of multicellularity (Buss 1987). Buss's major contribution was to highlight conflict suppression as the key to the origin and maintenance of a new level of individuality. In the case of multicellularity, Buss paid particular attention to the separation of the germline from somatic cells as an adaptation to prevent selfish cells, like cancer cells, from being passed onto the next generation. The emphasis on internal conflict suppression turned the problem of hierarchy into a problem of cooperation and brought it into the remit of social evolution.

At a conceptual level, Buss (1987) contrasted his hierarchical, multilevel selection approach with the gene-centred approach of Dawkins and Williams. Though he clearly came down on the side of the hierarchical approach, he stressed that the issue was primarily one of taste: 'to adopt a gene selection perspective is not wrong. It simply does not help unravel the central dilemma of our science' (Buss 1987, p. 55). The case for a gene-centred approach to hierarchy was taken up by Maynard Smith and Szathmáry. On this point they were unequivocal: 'we are committed to the gene-centred approach outlined by Williams (1966) and made still more explicit by Dawkins (1976)' (Maynard Smith and Szathmáry 1995, p. 8). The conflict between the two perspectives is, in some ways, exaggerated and several authors have argued that the two are compatible and complementary (including Queller 1997; Michod 1999; Okasha 2006; Bourke 2011). Not everyone, however, took kindly to the gene's-eye approach to hierarchy. A review of Maynard Smith and Szathmáry (1999) in *Nature* by Gabby Dover began:

One of the sadistic pleasures to be had from the defunct age of selfish-genery was to witness the mental loops of its proponents as they tried to extricate themselves from the illogical cul-de-sacs of their own devising. In his writings, Richard Dawkins pseudo-'paradox of the organism' was the climactic apotheosis of a belief in his own rhetorical devices which forced him to suspend all scientific rationale and modesty. (Dover 1999)

Dover went on to elaborate on this attack in his essay 'Anti-Dawkins' (Dover 2000). The 'paradox of the organism' that Dover refers to is the paradox that despite all the opportunity for within organism-conflict, the organism is 'not torn apart by the conflicting interests of the multitude of self-interested units that it contains' (Dawkins 1990). In fact, for many questions in evolutionary biology the consequences of within-individual conflicts can typically be ignored and the individual treated as a fitness maximizing agent. Dawkins was particularly interested in parasites manipulating the behaviour of their hosts as an example of extended phenotypes (see Chapter 5), but he clearly appreciated the fact that the integrated cohesion of individual organisms is not something that can be taken for granted, but is something that demands an evolutionary explanation.

Overall, Dawkins's own attitude to hierarchy appears somewhat ambivalent. In the early 1980s, when he had just taken on the editorship of the newly formed *Oxford Surveys in Evolutionary Biology* together with Mark Ridley, he wrote to the palaeontologist Niles Eldredge asking him to contribute a paper. At the time, Eldredge had recently proposed the theory of punctuated equilibria together with Stephen Jay Gould (Eldredge and Gould 1972) and was a major proponent of hierarchical approaches to evolutionary theory (though not those now associated with the major transitions). Eldredge replied that he did indeed have an appropriate manuscript. But, he added, 'you aren't going to like this, this is all about hierarchy'. Dawkins quickly wrote back saying 'What makes you think I don't like hierarchy?' Eldredge later said of the exchange: '[that was] a very amusing and witty thing for him to write, because he's such a gene oriented, reductionist person. But he does talk about hierarchies, he just handles them differently' (Eldredge 1996, p. 91).

In *The Selfish Gene*, Dawkins clearly presents the gene's-eye view as an empirical alternative to (Wynne-Edwards style) group selection. In *The Extended Phenotype*, this claim is toned down and, as Buss (1987, p. 187) and Okasha (2006, p. 222) have pointed out, he unwittingly engages in a bit of group selection theorizing himself in his discussion of how the first replicators came together to form the first genomes and cells. Dawkins clearly thought the origin of the first genome was an interesting problem and once implored Maynard Smith to spend more time developing game theory for replicators rather than individuals (Dawkins 1978). Dawkins's lack of clarity may well stem from him always considering the interactor question as secondary to the beneficiary question (*cui bono*).

On one point, though, Dawkins does not hesitate: genes are not the lowest level in the hierarchy. Right after the appearance of *The Selfish Gene*, Gould wrote:

> challenges to Darwin's focus on individuals have sparked some lively debates among evolutionists. These challenges have come from above and from below. From above, Scottish biologist V.C. Wynne-Edwards raised orthodox hackles fifteen years ago by arguing that groups, not individuals, are units of selection, at least for the evolution of social behavior. From below, English biologist Richard Dawkins has recently raised my hackles with his claim that genes themselves are units of selection, and individuals merely their temporary receptacles.   (Gould 1977)

In a similar vein, Elliott Sober and David Sloan Wilson argued: 'A proper understanding of the units of selection problem must take account of an important symmetry: *Just as organisms are parts of groups, so genes are parts of organisms*' (Sober and Wilson 1994; original emphasis). Dawkins would later admit that Gould's comparison of him to Wynne-Edwards tickled his 'sense of mischief' (Dawkins 1982b), but he also thought that Gould (and, by implication, Sober and Wilson) was confused about the difference between replicators and vehicles: '[Gould] keeps going on about hierarchy as though the gene is the bottom level in the hierarchy. The gene has nothing to do with the bottom level in the hierarchy. It's out to one side' (Dawkins 1996). To Dawkins, there is no symmetry comparing his genes to Wynne-Edwards's groups. The individual versus group dispute was a matter of empirical facts, falling under either Lloyd's interactor or manifestor of adaptation question. The gene, in contrast, was the answer to the beneficiary question. Williams also maintained that comparing genes and organisms leads you astray:

> Until you've made the distinction between information and matter, discussions of levels of selection will be muddled. Comparing a gene with an individual, for instance, in discussions of levels of selection, is inappropriate, if by 'individual' you mean a material object and by 'gene' you mean a package of information.   (Williams 1996b, p. 44)

To Dawkins and Williams, genes really are special. Again, this conclusion ties back to the type versus token distinction. The gene token is clearly at the bottom of the hierarchy and the only way to have the gene be on the side is to adopt the type perspective where genes are packages of information. Gould would later endorse Lloyd's resolution to the levels of selection question (Ketcham 2018), but the role of the gene's-eye view's replicator in major transitions remains contentious.

Some authors have argued that the major transitions are best approached using multilevel selection analysis (Buss 1987; Michod 1999). This has led the hierarchy to be compared to Russian matryoshka dolls with layers of conflict and cooperation nested within another (Wilson and Wilson 2008). An alternative to multilevel selection models is inclusive fitness analysis (Bourke 2011, 2014; West et al. 2015). What all approaches have in common is a commitment to search for common principles to answer the question that unifies all levels (Ågren et al. 2019a): why does natural selection favour cooperation rather than selfish behaviour that would undermine the integrity of higher levels?

The problem with the Russian doll image is that in a matryoshka doll there are dolls all the way down. Under the gene's-eye view, this is not how the biological hierarchy works (Bourke 2011, p. 59). Whereas vehicles can be said to make up other vehicles, in the way a multicellular organism is a group of cells, replicators (genes) play a different role. As Dawkins put it above, the replicator is out to one side (Dawkins 1996). Instead of Russian dolls, Bourke suggests that a more apt description would be the children's game 'pass the parcel'. In this game, a parcel is passed from child to child, each unwrapping one of the many layers of wrapping paper. At the centre, however, is not another piece of wrapping paper, but instead the gift, the parcel's actual purpose; or, to borrow Lloyd's terminology, the beneficiary of the whole unwrapping process.

In the biological hierarchy, genes are not just another level, but the actual beneficiary of selection at all levels. Andrew Bourke has explicitly argued in favour of a gene-centred inclusive fitness approach to social evolution in general and the major transitions in particular (Bourke 2011, 2014). In Chapter 4, I will examine the relationship between the gene's-eye view and inclusive fitness and argue that while the two are intimately linked, there is also an under-appreciated potential tensions between them.

2.6 Summary

- The gene of the gene's-eye view is agnostic about molecular details and is instead defined as a difference maker that is faithfully transmitted across generations. Ignoring molecular details works best when sorting out the logic of evolutionary scenarios with the goals of answering the so-called beneficiary question.
- This gene concept can be generalized to replicators, which are entities whose structure is passed on intact across generations. To fully capture the evolutionary process, replicators need to be complemented by vehicles, which are cohesive wholes that interact with the environment to determine reproductive success. In organic evolution, vehicles are typically played by individual organisms, and

more rarely by cells and groups. Distinguishing between replicators and vehicles turned out to be very helpful in moving the debate about the gene's-eye view forward.

- Memes were introduced to be a cultural replicator, equivalent to genes in organic evolution. Despite being a hugely successful meme in itself, memetics has had limited influence on the contemporary study of cultural evolution.

- While the replicator–vehicle distinction may be less general than the 'classic' recipe tradition of abstract formulations of natural selection, it has nevertheless been a very productive way to think about evolution.

- The emergence of the major transitions research programme revitalized the level-of-selection debate. While the gene's-eye view can fit comfortably within the major transitions framework, the latter has exposed further weaknesses in the replicator versus vehicle distinction.

3

Difficulties of The Theory

3.1 Introduction

The gene's-eye view has been under intense scrutiny ever since its conception. For one, the publication of *The Selfish Gene* only a year after E.O. Wilson's *Sociobiology* meant that it was immediately caught up in the sociobiology debate. Public scrimmages included the strange Isadore Nabi affair in the pages of *Nature* where a group of biologists (believed to have been Richard Lewontin, Richard Levins, Leigh van Valen, and Robert MacArthur) wrote under a pen name to criticize sociobiology, including the gene's-eye view (Nabi 1981a, 1981b; Lester 1981; Lewontin 1981; van Valen 1981; Wilson 1981; see Segerstrale 2000, pp. 184–188).

Over in *The New York Review of Books*, Stephen Jay Gould and Daniel Dennett had an ill-tempered exchange over the former's review of Helena Cronin's *The Ant and the Peacock* in 1993. The core thesis of Cronin's book was that the field of evolutionary biology had collectively settled that the gene's-eye view provided the key to two long-standing problems: the evolution of altruism and sexual selection, illustrated by the ant and the peacock respectively. Gould vehemently rejected this conclusion. Titling his review 'The confusion over evolution', he relegated gene selectionism to a 'marginal position among evolutionists' (Gould 1992). Maynard Smith, who had written the preface of Cronin's book, was surprised by Gould's tone and wrote in to say so (Maynard Smith 1993). So too did Dennett, though he

The Gene's-Eye View of Evolution. J. Arvid Ågren, Oxford University Press. © J. Arvid Ågren 2021.
DOI: 10.1093/oso/9780198862260.003.0004

turned the rhetoric up a notch (Dennett 1993). In his reply, Gould described Maynard Smith's and Dennett's letters as a 'good-cop–bad-cop grilling' (Gould 1993). The section of the reply dedicated to Maynard Smith began with a nod to 'My dear colleague and good-cop John Maynard Smith' and referred to 'the little community of professional evolutionists (that John and I proudly call our own)...' The part dedicated to Dennett, in contrast, opened with:

> The less than collegial tone of Daniel Dennett's commentary affirms the worst suspicions bruited in some quarters about the pungently rarified air of Cambridge, Massachusetts. (Thank God for Fenway Park and my local Bowl-a-Drome, where these mental pirouettes can be temporarily put aside and a semblance of populist normality attained.) Really, Dan, however much you may find my views on adaptation distasteful, why do you use this forum to air your personal grievances?
>
> (Gould 1993).

Disagreements over the gene's-eye view have often been heated. Not all critiques, however, were aired in such public venues and with such bellicose rhetoric. Many of the criticisms were fair and reflected interesting disagreements over how to think about evolution and natural selection. Others were less so. In this chapter, I focus on some 'difficulties of the theory', to borrow Darwin's phrase, that have received much attention.

The first concerns an old sin of biology: anthropomorphizing. The intentionality and personification involved in calling genes selfish has grated critics inside and outside of biology since day one. While calling genes selfish may seem innocent—'because no sane person thinks DNA molecules have conscious personalities', as Dawkins (2016, p. viii) put it—it does reflect a long-standing division within biology about the role of teleology (explanations in terms of purpose and final causes) and intentional language. Such language is common in evolutionary biology but is viewed with deep scepticism by other biologists. I discuss this disagreement and argue that despite its weaknesses, intentional language has an important role to play in biology.

The second concerns how the gene's-eye view handles interactions between genes. In *The Selfish Gene*, Dawkins develops an analogy between genes in an organism and oarsmen in a rowing crew. He argues that even if the coach is only basing their selection on how successful each crew is, the coach is still selecting for the best individual rowers. Critics argue that this is misleading. I use the rowing example to discuss how the gene's-eye view handles epistasis and heterozygous advantage, and whether it commits the so-called averaging fallacy.

The third concerns the charge that the gene's-eye view confuses bookkeeping with causality. This is a long-standing criticism that holds that just because the evolutionary outcome can be described as a change in allele frequencies, it does not follow that causality is best assigned to the level of genes. While biologists such as Stephen Jay Gould popularized this criticism, much of the debate surrounding it played out in the philosophical literature. I review the debate, showing how it provides an informative account of the concept of pluralism in evolutionary explanations.

The fourth concerns genetic determinism. This is the assertion that the gene's-eye view allows genes a too privileged role in accounts of development. According to critics, genes play no special causal role and biology needs a more inclusive notion of inheritance. The most ambitious version of this claim has been put forward by developmental system theorists, in the light of which I evaluate where this leaves the gene's-eye view today.

The fifth concerns the concept of human nature and what evolutionary theory can tell us about human affairs. Evolutionary biology in general has had a troubled relationship with both, and the notion of selfish genes struck at the heart of it. This debate has recently calmed down, but there are interesting parallels between the views of George Williams and Richard Dawkins, and Darwin's contemporary T.H. Huxley.

3.2 Anthropomorphizing

In 1979 the English philosopher Mary Midgley penned 'Gene juggling', an infamous commentary on *The Selfish Gene* that began: 'Genes cannot be selfish or unselfish, any more than atoms can be jealous, elephants abstract or biscuits teleological' (Midgley 1979). 'Gene juggling' was not a book review, but a response to a paper on reciprocal altruism that drew heavily on *The Selfish Gene* by another philosopher, John Mackie (Mackie 1978). Still, Midgley's tone was unusually harsh and, perhaps unsurprisingly, Dawkins's reply was cutting. In 'In defence of selfish genes', he deploys his full stylistic repertoire (Dawkins 1981a). It begins:

> I have been taken aback by the inexplicable hostility of Mary Midgley's assault. Some colleagues have advised me that such transparent spite is best ignored, but others warn that the venomous tone of her article may conceal the errors in its content. Indeed, we are in danger of assuming that nobody would dare to be so rude without taking the elementary precaution of being right in what she said.
>
> (Dawkins 1981a)

Dawkins goes on to describe how Midgley 'raises the art of misunderstanding to dizzy heights' and lambasts her for her focus on humans, which Dawkins admits being rather uninterested in ('a particular, rather aberrant species'). Midgley operated under the impression that Dawkins actually thought that genes were selfish in the same way as humans are, rather than in the more technical way used by evolutionary biologists. An exasperated Dawkins asked: 'Did Midgley, perhaps, just overlook my definition? One cannot, after all, be expected to read every single word of a book whose author one wishes to insult' (Dawkins 1981a).

Midgley's paper represented an especially spurious misunderstanding of the gene's-eye view. In fairness, Midgley did later apologize for the 'impatient tone' of her article (Midgley 1983). She would go on to develop a more nuanced take in her subsequent writings, such as *Evolution as Religion* (Midgley 1985) and *The Solitary Self: Darwin and the Selfish Gene* (Midgley 2010), though still viewing *The Selfish Gene* as 'a rotten essay in moral philosophy, propped up with bad scientific examples' (quoted in Brown 2016).

While this particular misunderstanding has never been particularly widespread (but see, e.g., Stove 1992 and Stove 1995 for an effort to keep the idea alive), the habit of ascribing goals, aims, and strategies to individual genes tends to annoy certain kinds of biologists and philosophers alike. For example, Rosenberg, who otherwise has been very supportive of the gene's-eye view, describes biologists using this kind of language as 'conspiracy theorists' (Rosenberg 2011, pp. 13–14) and Francis (2004, p. 8) and Godfrey-Smith (2009, p. 10) as suffering from 'Darwinian paranoia'. The plant scientist David Hanke diagnosed the state of biology as follows:

> Biology is sick. Fundamentally unscientific modes of thought are increasingly accepted, and dominate the way the subject is explained to the next generation. The heart of the problem is that we persist in making (literal) sense of a world that we now know to be senseless by attributing subjective values to the objectives in it, values that have no basis in reality.   (Hanke 2004)

Hanke contrasts his own scientific training, which taught him 'to reason critically and objectively (...) and to regard the subjective as untrustworthy and deceiving, potentially corrupting the truth' with the 'tragedy' that his undergraduate students have read nothing but Dawkins before showing up in his classroom (Hanke 2004). The students were all apparent victims of what Lucy Sullivan (1995) called the 'Oxford school of biological science fiction'. To Hanke and Sullivan, the most egregious fault of anthropomorphizing, and intentional language more broadly, is not just that it is 'lazy and wrong' (Hanke 2004) and whose 'hegemony [is] quite out of proportion to its intellectual finesse' (Sullivan 1995), but that it leads us astray when confronted with biological problems.

Disagreements about the role of anthropomorphizing, teleology, and intentional language did not start with *The Selfish Gene* (Ruse 1989). Instead, Hanke's and Sullivan's critique echoes a general argument that such language is an embarrassment that makes biology look like an immature science next to the likes of physics and chemistry. In those subjects, they argue, theories are strictly physical and mechanical, with all talk of purpose, goals, and intentions banished. Why should biology be different?

This is in many ways a sensible question. Throughout the history of biology, advances were often made by scientists who did not consider biology to be anything special. For example, in the 17th century the French philosopher René Descartes argued that living organisms were nothing but machines that could be understood accordingly. This quickly became the prevalent attitude, and many new discoveries were made treating parts of organisms as parts of machines and then asking how they all fit together (see Riskin 2016).

Another example of the non-uniqueness of biology comes from Erwin Schrödinger *What is Life?* lectures delivered at Trinity College Dublin in 1944. At that time the material basis of heredity was not yet known. Part of the transformative value of the lectures laid in presenting the problem of heredity as a physio-chemical problem like any other. This would end up having a profound effect on the audience and those who read the book subsequently published with the same title. Two such young readers were Francis Crick and James Watson who, together with Maurice Wilkins and Rosalind Franklin, would later decipher the double helical structure of DNA. Both credited Schrödinger for switching their research from physics and ornithology, respectively, to molecular biology.

The discovery of the double helix in 1953 ushered in a golden age of molecular biology. Associated with this was an influx of people and methods from physics and chemistry. More recently, the flood of data produced by whole genome sequencing and other technological advances has attracted computer scientists, engineers, and statisticians to the life sciences. This shift in biological research takes us back to the question above, why should biology be different from other sciences? Or, in other words, what, if anything, makes biology unique as an autonomous scientific discipline?

The simplest answer is that biology is about organisms and organism are special. Organisms are material things, made of the same fermions and bosons as icebergs, and subject to the same physical laws as candle flames. At the same time, they are unlike any other material thing. Organisms appear to be endowed with a goal-directedness absent in non-living things (Wilson 2005; Walsh 2015; Ruse 2018). It is therefore in biology that purpose enters into scientific explanations. That organisms make biology unique was also the answer Immanuel Kant arrived at after much deliberation (Huneman 2007). To Kant, it was impossible to talk about biology without using purposeful language. Kant did not believe that plants and animals actually possessed some teleological or purposeful force. Instead, purpose is the most powerful heuristic we have to talk about living things. Kant's conclusion left him rather disappointed and he glumly concluded that biology could never be as proper a science as physics:

> it is absurd to hope that another Newton will arise in the future who will make comprehensible to us the production of a blade of grass according to natural laws that no design has ordered. Such insight we must absolutely deny to mankind.   (Kant 1790, p. 228)

All of this changed with Darwin. In his theory of evolution by natural selection, he provided an account of how a purely mechanistic process can lead to the appearance of design in nature (Dennett 1995; Haig 2020). Evolution by natural selection provides the bridge between mechanism and purpose.

3.2.1 Reading Mother Nature's mind and licensed anthropomorphism

Although it can be appropriate to talk about purpose, to think in terms of intentions, to anthropomorphize, this does not mean that all ways of doing so are equally good. Like all heuristics, the goal of anthropomorphizing is to make sense of old data and to generate new empirically testable hypothesis about biological phenomena. Saying that mosquitos bite because they dislike you, or that moths fly into flames because they are depressed by the current state of political affairs, do neither. That kind of naive anthropomorphizing is not helpful.

Biologists have typically used two types of anthropomorphizing, more accurately referred to as 'agential thinking' (coined by Godfrey

Smith 2009 and elaborated by Okasha 2018). The first conceptualizes the process of natural selection itself as an agent trying to choose the best individuals to survive and reproduce, analogous to a farmer choosing the best stallion for breeding. This tradition, which has been referred to as 'reading Mother Nature's mind' (Okasha 2018, p. 21), goes back to Darwin himself who spoke of natural selection as 'rejecting that which is bad' and 'preserving and adding up all that is good' (Darwin 1859, p. 83). He also made much of the parallels between artificial and natural selection in *The Origin of Species*:

> Under nature, the slightest difference of structure or constitution may well turn the nicely-balanced scale in the struggle for life, and so be preserved. How fleeting are the wishes and efforts of man! how short his time! and consequently how poor will his products be, compared with those accumulated by nature during whole geological periods. Can we wonder, then, that nature's productions should be far 'truer' in character than man's productions; that they should be infinitely better adapted to the most complex conditions of life, and should plainly bear the stamp of far higher workmanship? (Darwin 1859, pp. 83–84)

In retrospect, the use of the phrase 'far higher workmanship' is rather Paleyan (Ruse 2019b, p. 39). Despite its distinguished origin, this kind of thinking has some serious limitations. In particular, it may lead us to believe that evolution has a goal and that natural selection acts with foresight, both of which are clearly false.

In contrast, the second type of agential thinking has a lot going for it. It conceives of biological entities, such as genes, cells, or organisms, as agents pursuing goals. In contemporary evolutionary biology, this approach is best represented by proponents of inclusive fitness who argue that organisms should appear as-if designed to maximize their genetic representation in future generations, which can be captured by modelling their maximization of inclusive fitness (Grafen 2007, 2014a; Gardner 2009; West and Gardner 2013). Agential thinking at the organismal level has proven to be especially popular in behavioural ecology (see, e.g, Davies et al. 2012 for a textbook with it as its conceptual backbone).

Assigning agency to genes may at first seem strange. Agency has often been considered a defining distinction between the living and the non-living (see, e.g., Soto and Sonnenschein 2020) and the gene's-eye view transfers agency from the paradigm living object—organisms—to something non-living, genes. Moving beyond this initial discomfort, agential thinking at the gene level finds its rationale in helping biologists think about difficult problems. The gene's-eye view helped make sense of old problems (Cronin 1991) and opened our eyes to new ones (Chapter 5).

When assigning agency to genes, Dawkins repeatedly warned readers to be cautious. With little success, it seems: '[Dawkins's] caveats slowed readers down about as effectively as "Slow—Work Zone" signs on a deserted highway', as Nesse (2006) put it. The importance of caution was also at the heart of Dawkins's reply to Lucy Sullivan's diatribe against anthropomorphizing in *The Selfish Gene* (Sullivant 1995; Dawkins 1995a). Dawkins originally made the point as follows:

> If we allow ourselves the license of talking about genes as if they had conscious aims, always reassuring ourselves that we could translate our sloppy language back into respectable terms if we wanted to, we can ask the question, what is a single selfish gene trying to do?
>
> (Dawkins 1976, p. 88)

What does 'respectable terms' mean? John Maynard Smith, when pressed on this point in a Q&A-session following one of his talks, answered:

> I am prepared to think as loosely as necessary to give me an idea when I'm confronted with a new biological problem. If it helps me think to say, 'If I was a gene, I would do so-and-so', then I think that is OK. But when I've got an idea, I want to be able to write down the equations and show that the idea works. (...) I'm all for loose thinking. We all need ideas.   (Maynard Smith 1998)

Dawkins and Maynard Smith emphasized the importance of both having tools to support themselves as biologists thinking through difficult problems and then translating their insights into a more respectable form. The balance between anthropomorphic 'loose thinking' and more formal reasoning, usually provided by mathematics, is nicely captured by Grafen's term 'licensed anthropomorphism' (Grafen 2003; though he coined it to refer to the idea that Fisher's Fundamental Theorem gives the license to think of organisms as fitness maximizers). Licensed anthropomorphism comes close to my own preferences in using a gene's-eye view heuristic, coupled with a simple population genetic model to show that the logic holds (such as in Ågren et al. 2019b).

A good illustration of the value of licensed anthropomorphism comes from the work of Manus Patten. In a recent paper, Patten asked the simple question: should a mutation that arises on the X-chromosome favour males or females? From a gene's-eye view, it can be predicted that because a X-linked gene spends two thirds of its time in females, and only one third of its time in males (assuming an XY sex chromosome system and a 1:1 sex ratio), that it generally should favour females (Frank and Crespi 2011). Using a simple population genetic model, however, Patten showed that you can get the opposite result: mutations that favour males at the expense of females can more readily spread in a population (Patten 2019). This comes about because in males each X-linked gene is present in only one copy, and deleterious effects therefore cannot not be masked by the other allele. Patten and Frank later synthesized their models to show that we should expect a mosaic of both male and female biased mutations on the X (Frank and Patten 2020; see also Hitchcock and Gardner 2020). The exact distribution of the two will depend on assumptions about dominance, dosage compensation, and the magnitude of sexually antagonistic selection.

3.3 Epistasis, heterozygote advantage, and the averaging fallacy

One of the most famous analogies in *The Selfish Gene* is that of the genome as a rowing boat and genes as oarsmen. The analogy starts with the Boat Race that takes place annually between Putney and Mortlake along the River Thames in London between the universities of Oxford and Cambridge. Each team is made up of nine individuals, eight rowers, and one cox who steers the boat. Winning requires oarsmen who are not only individually accomplished but also capable of meshing as a team. How, then, should the coach select the team to balance individual and team skills? Dawkins suggests the following strategy:

> Every day [the coach] puts together three new trial crews, by random shuffling of the candidates, for each position, and he makes the three crews race against each other. After some weeks of this it will start to emerge that the winning boat often tends to contain the same individual men. These are marked up as good oarsmen. Other individuals seem consistently to be found in slower crews, and these are eventually rejected. But even an outstandingly good oarsman might sometimes be a member of a slow crew, either because of the inferiority of the other members, or because of bad luck—say a strong adverse wind. It is only on average that the best men tend to be in the winning boat.
>
> (Dawkins 1976, p. 40)

Moving back to genes. Competitors for each seat on the boat correspond to alleles and rowing the boat to victory equals building an organism with high fitness. In Chapter 1, I outlined how the gene's-eye view leads to an expanded definition of environment, one that also includes non-additive genetic effects such as dominance and epistasis. Selection will then favour genes that do well in the environment provided by other genes. As Williams put it in *Adaptation and Natural Selection*:

> No matter how functionally dependent a gene may be, and no matter how complicated its interactions with other genes and environmental factors, it must always be true that a given gene substitution will have an arithmetic mean effect on fitness in any population. (...) Adaptation can thus be attributed to the effect of selection acting independently at each locus.   (Williams 1966, pp. 56–57).

When reading Williams's and Dawkins's writing above, the influence of Fisher (1918), as well as Fisher (1941), really shine through. Even so, in an interview with Ullica Segerstrale, Lewontin described the above averaging strategy as 'an epistemological error' stemming from 'a lack of understanding of population genetics' (quoted in Segerstrale 2000, p. 277).

There are several points to disentangle here. One is about how population size may affect the fate of alleles. Take Dawkins's comment that appears after the oarsmen analogy is explained: 'a gene which is *consistently* on the losing side is not unlucky; it is a bad gene' (Dawkins 1976, p. 41; original emphasis). This addition is a little misleading, as it would only hold true in a scenario where the population size is infinite (or at least very, very large) and genetic drift is absent. That is clearly unrealistic. Many features of genome evolution and structure are likely due to genetic drift (Lynch 2007a).

A gene may also be unlucky because of population structure. For example, if the trial crews in Dawkins's example are not put together at random, but instead certain oarsmen are more likely to be grouped together than others, then the coach's averaging strategy will not work (Bar-Yam 1999). In general, this averaging strategy poses a serious problem for the gene's-eye view. In many situations, there will be way more possible genotypic combinations than there are atoms in the universe. While approximations will be of some help, getting the true average effect of an allele on a phenotype will be possible only in controlled lab experiments, not in most natural populations.

An additional issue is the question of epistasis. Epistasis is another one of those terms that can lead to frustration when molecular and evolutionary geneticists interact, as it refers to different biological phenomena in the two fields (Wade 1992). In molecular genetics, epistasis happens when a mutation in one gene masks the expression of another gene, as when a dominant allele masks the effect of a recessive allele. In evolutionary genetics, however, it refers to the non-additive component of genetic variance, which results from interactions among different loci. In this particular debate, everyone is talking about epistasis in the latter, statistical sense.

3.3.1 Ernst Mayr, Sewall Wright, and the matter with epistasis

Echoing his critique of 'beanbag genetics', Mayr expressed the epistasis criticism of gene-centred approaches to evolution as follows: 'no gene has a fixed selective value, the same gene may confer high fitness on one genetic background and be virtually lethal on another' (Mayr 1963, p. 296). To him, it made little sense to talk of the effect of individual genes. He therefore repeatedly emphasized the concept of 'unity of the genotype' and the importance of 'balanced complexes that resist change' (Mayr 1975; see Mayr 1997 for a summary of his views on the issue).

As is often the case in the history of evolutionary genetics, the disagreement over epistasis can be introduced as a disagreement between Sewall Wright and R.A. Fisher. Wright and Fisher had different views on many things, but the role of epistasis in adaptive evolution was at the heart of one of their fiercer disagreements (Provine 1986, Chapter 8). In his review of Fisher's *The Genetical Theory of Natural Selection*, Wright wrote that the two of them would be:

> in exact agreement in all cases only if dominance and epistatic relationships were completely lacking. Actually, dominance is very common and with respect to such a character as fitness, it may safely be assumed that there are always important epistatic effects. Genes favorable in one combination, are, for example, extremely likely to be unfavorable in another.   (Wright 1930).

Wright reiterated this view many times throughout his career, including near the very end when, in his nineties, he used *The Selfish Gene* to introduce a paper in *Evolution* (Wright 1980). Wright, as opposed to Fisher and Haldane, was thus the only one of the three founders of population genetics who lived long enough to comment directly on the gene's-eye view debate. While the paper primarily presents Wright's general views on the state of population genetics and his contribution to the field, the paper was also intended to be a rejection of the gene's-eye view, which is how it has usually been read. Yet, the paper itself touches only briefly on Williams, Dawkins, and Maynard Smith, whom Wright lists as contemporary representatives of Fisher's views. Wright must therefore have wondered what to make of Peter Medawar's comment in the *New York Review of Books* that:

> The most important single innovation in the modern synthesis was however the new conception that a population that was deemed to undergo evolution could best be thought of as a population of fundamental replicating units—of genes—rather than as a population of individual animals or of cells. Sewall Wright (...) was a principal innovator in this new way of thinking—a priority for which R.A. Fisher, an important but lesser figure, never forgave him... (Medawar 1981)

Dawkins, however, admitted to 'enthusiastically endorsing almost every word' of Mayr (1963) and out of all of Wright's (1980) arguments he disagreed with 'almost none' (Dawkins 1982a, p. 238). Furthermore, Dawkins argued that when the gene's-eye view is properly understood, it leads to an emphasis on genes' ability to cooperate and indeed can offer not only a version of Wright and Mayr's views that they may find palatable, but also a 'truer and clearer expression' of them (Dawkins 1982a, p. 239).

Dawkins returned to the rowing example later in *The Selfish Gene* and discussed what would happen to the speed of the boat if the oarsmen communicated with each other (Dawkins 1976, pp. 91–92). Imagining a scenario where all oarsmen are monolingual, half English speakers and half German speakers, Dawkins argued that the end-result would be crews of only English speakers or only German speakers. The absence of mixed crews, however, does not mean that the coach selected on the crew as a unit. The coach was still selecting based on individual skill, but that central to that skill was the ability to communicate with the other oarsmen in the same language. In *Unweaving the Rainbow,* he went on to devote a full chapter, titled 'The selfish cooperator', to the issue of genes cooperating (Dawkins 1998b, pp. 210–234). He has also suggested that *The Cooperative Gene* would have made for an equally good title as *The Selfish Gene* (a point with which his editor Michael Rodgers strongly disagrees with to this day; M. Rodgers personal communication).

Opinions differ on how successful Dawkins was with this explanation. Wade reviewed the mathematics of epistasis in light of the gene's-eye and suggested that it usually provides a too simplistic account of the underlying genetic interactions (Wade 2002; see also Kamath 2009). Similarly, Gould argued that the gene's-eye view works 'only if organisms developed no emergent properties—that is, if genes built organisms in an entirely additive fashion, with no non-linear interaction among genes at all, [but] organisms are stuffed full of emergent properties' (Gould 2002, p. 620). Discussing an earlier version of this argument (laid out in Gould 1977), Dawkins maintained that Gould confused genetics and embryology: 'if this were really a good argument, it would be an argument against the whole of Mendelian genetics, just as much as against the idea of the gene as the unit of selection' (Dawkins 1982a, p. 116). Dawkins noted that because genes are inherited as discrete objects, any 'blending' (that is non-linear interactions between genes) that may happen during development is irrelevant.

Dawkins's and Gould's disagreement is partly a reflection of their different uses of the word 'gene.' As outlined in Chapter 2, the gene's-eye view's abstract conception of a gene as a stably inherited difference maker is quite different from the empirically informed definition preferred by Gould. The weakness of Dawkins's argument is that just because genes are stably inherited it is not necessarily reasonable to assume that that is the case for their 'difference making' qualities. Dawkins's Mendel-point holds under his preferred definition but misses the point in light of alternative definitions.

3.3.2 Heterozygote advantage and the averaging fallacy

Sober and Lewontin's paper 'Artifact, cause and genic selection' (Sober and Lewontin 1982; see also Sober 1984, pp. 302–313) is one of the most influential critiques of the gene's-eye view. Even Maynard Smith admitted to having been initially swayed by the paper, though in the end he remained unconvinced (Maynard Smith 1987). Sober and Lewontin's argument was that while it is always possible to calculate selection coefficients for individual genes, those are not actually real; they are statistical artifacts.

To illustrate this point, they developed an example of heterozygote advantage. Heterozygote advantage arises when the heterozygous genotype (say, Aa) has a higher fitness than either of homozygous genotype (AA or aa). The textbook example of heterozygote advantage is the balancing selection acting on the gene involved in sickle-cell anaemia and malaria resistance. Individuals homozygous for the recessive allele develop sickle-cell, whereas heterozygotes are resistant to malaria, resulting in both alleles being retained in populations with high prevalence of malaria.

Sober and Lewontin used a simple one-locus two-allele model to present their case. Consider the two alleles, A and a with no dominance, that have frequencies p and q respectively (such that $p + q = 1$). The population is assumed to be at Hardy–Weinberg equilibrium prior to selection, and the fitnesses of $AA, Aa,$ and aa are given by w_{AA}, w_{Aa}, and w_{aa}. The genotype frequencies after selection can be calculated by multiplying the initial genotype frequencies with their fitness and then dividing by the population mean fitness, $\bar{w}$ (Table 3.1).

So far, fitness is a property of the diploid genotype, not of individual alleles. Deriving the fitnesses of individual alleles (A and a), however, is not particularly complicated. For example, if the fitness of A is defined as W_A, it can be calculated by averaging the fitness of A across all genotypes in which it is found. Recall that A can occur with a frequency of 1, $1/2$, and 0 in $AA, Aa,$ and aa respectively. Then, using the frequencies and fitnesses defined above, and given that: the

Table 3.1. Sober and Lewontin's heterozygote advantage example

Genotype	AA	Aa	aa
Frequency before selection	p^2	$2pq$	q^2
Fitness	w_{AA}	w_{Aa}	w_{aa}
Frequency after selection	$\dfrac{p^2 w_{AA}}{\bar{w}}$	$\dfrac{2pq w_{Aa}}{\bar{w}}$	$\dfrac{q^2 w_{aa}}{\bar{w}}$

frequency of A after selection $\times \bar{w} = w_A \times$ frequency of A before selection, this gives:

$$w_A = w_{AA}p^2 + w_{Aa}((1/2)2pq)/(p^2 + (1/2)2pq) = pw_{AA} + qw_{Aa}$$

Using the same method:

$$w_a = w_{aa}q + w_{Aa}p.$$

w_A and w_a can then be used to predict changes in allele frequency.

Sober and Lewontin granted that this will work most of the time. With heterozygote advantage, where $w_{Aa} > w_{AA}$ and $w_{Aa} > w_{aa}$, however, they argue that the approach fails. While individual selection coefficients can readily be calculated in this case by averaging across all genotypes, and both the individual gene model and the diploid genotype model can predict future changes in gene frequencies, only the latter accurately represents the causal forces at play.

To Sober and Lewontin, the problem is that A lacks what they call a 'uniform causal role'. In certain contexts it is favoured by selection and in others it is not: 'If we wish to talk about selection for a single gene, then there must be such a thing as the causal upshot of possessing that gene' (Sober and Lewontin 1982). Sober and Lewontin's argument has later been referred to as the 'averaging fallacy' (Okasha 2004), using a term coined by Sober and Wilson (1998, pp. 31–35). Sober and Wilson originally used the term in the group selection debate to criticize the suggestion that examples of group selection could be recast as individual selection by calculating the fitness of an individual by averaging its fitness across all the groups in which it is found. (While the gene versus genotype situation may appear analogous to the individual versus group one, there are crucial ways in which this analogy fails, Okasha 2004; but see Winther et al. 2013 for an alternative interpretation).

Several authors have defended calculating genic selection coefficients (including Rosenberg 1983; Maynard Smith 1987; Sterelny and Kitcher 1988; Waters 1991; Weinberger 2011; see also Queller 2020). The strongest defence is to concede that in the heterozygote advantage example the allele's effect is indeed context-dependent but to deny that this is a problem. Many traits are context-dependent and as such their fitness effects depend on their environment, which includes the states of the other traits in the organisms in which they are found. As explained above, under the gene's-eye view the rest of the genome, including the other allele in a diploid, is part of the environment. According to the gene's-eye view, then, the heterozygote case is no different from any other environment-induced variance in the fitness effect of an allele. While the gene's-eye view may work best if the effect of an allele is, or is close to being, independent of context (Okasha 2006, p. 167), imposing a requirement of a 'uniform causal role' is too stringent a condition.

3.4 The bookkeeping objection

In 2002, the year that he passed away, Stephen Jay Gould published *The Structure of Evolutionary Theory*. A majestic book (it clocks in at close to 1,500 pages) its vast scope touches on practically every aspects of evolutionary biology, its history, and its philosophy. Joe Felsenstein once remarked that 'if you're holding Gould's book in your lap, stasis is very real: you couldn't possibly stand up under the weight of all that verbiage' (Felsenstein 2012). In a similar vein, Stephen Stearns recalls bringing the book on a flight only to be asked by an air stewardess to store the book in the overhead compartment so not to risk hurting other passengers should he lose control of it during take-off and landing (Stearns 2002).

As expected, a significant chunk of the book is dedicated to explaining the perceived short-comings of the gene's-eye view. In a characteristically long-winded section, Gould writes that 'the error and the incoherence of gene selectionism (...) can be summarized in a single statement (...) proponents of gene selectionism have *confused bookkeeping with causality*' (Gould 2002, p. 632; original emphasis). The bookkeeping objection to the gene's-eye view states that while evolutionary change can be described as changes in allele frequencies over time, one cannot necessarily conclude anything about the causes of those changes. Just because the result of selection can be correctly recorded at the gene level, it does not mean that it is the most appropriate level of analysis. As Gould put it 'selection is a causal process, not a calculus of results' (Gould 2002, p. 623).

Whereas the credit for coining the term 'bookkeeping' belongs to Wimsatt (1980), it was advanced by several critics of the gene's-eye view (Sober and Lewontin 1982, 1983; Walsh 2004; Brosius 2005; Brandon and Nijhout 2006). A version of this argument can also be found already in Wright's criticism of Fisher (Wright 1930). Sober and Lewontin (1982) wrote that the genes'-eye view not only 'fails to do justice to standard textbook examples of Darwinian selection, it 'distorts the causal process', and simply 'generates the wrong answers to the question of what is happening'. The wrong answer that Sober and Lewontin have in mind is the gene's-eye view's desire to assign causality to the level of genes for all selective events. To them, the level at which causality belongs—gene, cell, individual, group—depends the biological example under study, and the gene level should be reserved for examples of genomic conflicts (see Chapter 5).

Gould probably did more than anyone to popularize the bookkeeping objection (Gould 1994, 2002, pp. 632–637, Gould and Lloyd 1999). Gould's version of the argument starts by conceding that genes are (in some ways) special: the faithful transmission of genes makes them a uniquely suitable entity to document the evolutionary history of a population. On this, he fully agrees with proponents of the gene's-eye view. Furthermore, because genes are at the very bottom of the biological hierarchy (genes within cells, cells within organisms, organism within groups, and so on) they possess an asymmetrical quality relative to other levels. That is, selection at higher levels also results in selection at lower levels, which means that if frequencies of higher levels shift, so do those of the lower levels. The reverse is not necessarily true. Selection at the genic level, for example on selfish genetic elements (see Chapter 5), may cause certain genes to increase in frequency

without having an effect on the species level (i.e. on the rate of speciation or extinction). Genes are thus the only level that can be said to
capture selection at all levels.

But, and to Gould this is the crucial but, this does not mean that
genes are the causal agents. Gould has argued that Williams and Dawkins
both realized this, but that they refused to admit it (Gould 2002,
pp. 632–644). Take, for example, Williams in *Natural Selection*:

> For natural selection to occur and be a factor in evolution, replicators
> must manifest themselves in interactors, the concrete realities that con-
> front a biologist. The truth and usefulness of a biological theory must
> be evaluated on the basis of its success in explaining and predicting
> material phenomena. It is equally true that replicators (codices) are a
> concept of great interest and usefulness and must be considered with
> great care for any formal theory of evolution.   (Williams 1992, p. 13)

And Dawkins in *The Extended Phenotype*:

> *Of course* genes are not directly visible to selection. *Obviously* they are
> selected by virtue of their phenotypic effects, and certainly they can
> only be said to *have* phenotypic effects in concert with hundreds of
> other genes.   (Dawkins 1982a, p. 117; original emphasis).

Such reasoning Gould takes as evidence that even the most ardent
advocates for the gene's-eye view actually agree with him: causality
belongs at the level of the interactors. Gould also cites the shift in
focus between *The Selfish Gene* and *The Extended Phenotype* as a case
in point. Early on, Dawkins clearly presents gene-level selection as a
(superior) alternative to group selection. For example, two years after
The Selfish Gene Dawkins wrote: 'Evolutionary models, whether they
call themselves group-selectionists or individual-selectionists, are fundamentally gene-selectionists' (Dawkins 1978). This is also the message
in *Adaptation and Natural Selection* (Williams 1966, pp. 123–124). In this
formulation, the gene's-eye view is meant to be a uniquely true
description of how the world works.

The argument in *The Extended Phenotype* is different. In the first
chapter of the book, Dawkins introduces the Necker cube illusion.

Under this illusion, the human mind can see two images of a two-dimensional line drawing of a cube, each equally 'true'. The gene's-eye view and the traditional individual-centred view are thus different, but equally valid, perspectives of evolution, just like the orientations of the Necker cube. To Gould, this sounds like George Aiken, the Republication senator from Vermont, who in the middle of the American war in Vietnam, suggested that the US should just declare victory and go home (Gould 2002, p. 638). Gould argues that Dawkins admits defeat, the gene is not the one true unit of selection, but that he still declares victory for the gene's-eye view.

The disagreement over the bookkeeping objection reflects a common confusion over whether gene selection is best regarded as a process or a perspective. This is a confusion for which admittedly Williams and Dawkins bear some responsibility. Dawkins argues that the group versus individual selection debate is simply an empirical disagreement over at what level selection is most effective in natural populations (Dawkins 1982b). The replicator versus vehicle distinction, on the other hand, is not. It is a question of perspective:

> I am not saying that the selfish organism view is necessarily wrong, but my argument, in its strong form, is that it is looking at the matter the wrong way up. (...) I am pretty confident that to look at life in terms of genetic replicators preserving themselves by means of their extended phenotypes is at least as satisfactory as to look at it in terms of selfish organisms maximizing their inclusive fitness. In many cases the two ways of looking at life will, indeed, be equivalent.
>
> (Dawkins 1982a, pp. 6–7)

So the replicator and vehicle perspectives may be equivalent. Does this mean that Dawkins considers them to be equally useful? The answer is clearly no. Remember that he said he coined the vehicle concept 'not to praise it, but to bury it' (Dawkins 1994b). Indeed, a central goal of *The Extended Phenotype* is to demonstrate the 'theoretical dangers' of thinking in terms of individual fitness (Dawkins 1982a, p. 91).

Gould's Aiken comparison is therefore rather uncharitable. Williams and Dawkins never denied that vehicles are often necessary for evolution (though not always, see Dawkins 1994b). Again, Lloyd's four questions (Chapter 2) are helpful to make sense of the situation. To Williams and Dawkins, the unit of selection question was never about what entities play the role of interactors; instead they were interested in Lloyd's beneficiary of selection question.

3.4.1 Tempered realism and pluralistic gene selectionism

To many biologists, this is where the debate ended. While the field of evolutionary biology has long wrestled with the concept of causality (see, e.g., Otsuka 2016 and Uller and Laland 2019 for overviews), many biologists, being primarily concerned with more practical matters, seem to have been quite happy to sometimes think of selection in terms of genes and sometimes in terms of individuals. The next phase of the debate over the bookkeeping objection therefore took place primarily in the philosophical literature. There it fed into a debate regarding whether a given selective event can always be causally described in one, and only one, way or in several, equally true, ways.

The idea of pluralism regarding selection is captured by Dawkins's Necker cube idea, that for any selection processes, there are multiple equally adequate representations of the causal forces at work. 'We have two images of natural selection', as Kim Sterelny and Philip Kitcher, put it, referencing the organism- and gene-centred accounts (Sterelny and Kitcher 1988). Together with Kenneth Waters's PhD thesis *Models of Natural Selection from Darwin to Dawkins* (Waters 1986), their paper marked the start of the debate over what has become known as pluralistic genic selectionism. Pluralism here contrasts with monism, the idea that there is a uniquely correct way to causally describe a selective event (Rosenberg 1993 makes this case). Genic pluralism considers the search for the 'true' unit of selection to be a waste of time, an 'exercise in muddled metaphysics' (Kitcher et al. 1990) and they urge their colleagues to 'turn their attention to more serious projects than that of quibbling about the real unit of selection'.

Kenneth Waters has advocated the term 'tempered realism' to describe this position (Waters 1991, 2005). The term comes about because the pluralists are still committed to scientific realism, the belief that science makes progress at describing true features of the universe, but that these descriptions are not uniquely true. That is, there may be multiple true causal models for a given biological event, hence the need to 'temper' the commitment to realism. For philosophers interested in causality this has been difficult to swallow. Some of the strongest critics of genic pluralism have been Sober (1990) and Lloyd (2005), and the latter's exchange with Waters (Waters 2005; Lloyd et al. 2005) reveals a debate that is very much still alive (see Ketchum 2018 for a summary).

If different models make the same empirical prediction, can, as Maynard Smith once put it, any good Popperian consider the matter of choosing between models as anything other than pseudoscience? (Maynard Smith 1987). There will be cases where one modelling framework is genuinely better than another at capturing the causal forces at work, but in others arguing over which one is correct would be 'as foolish as it would be to argue whether algebra or geometry is the correct way to solve problems in science' (Maynard Smith 1986, p. 30). That being said, the equivalence of models is often achieved only by comparing highly abstract statistical formalisms, rather than models that are explicitly casual. This, for example, is a problem with the claim of equivalence between inclusive fitness and group selection models (Clarke 2018; Birch 2019). There is therefore a serious discussion to be had about different modelling traditions in evolutionary biology, not just why we prefer one framework over another, but also what philosophical assumptions that lead us there (Otsuka 2019).

In a commentary on a paper that compared contextual and collective approaches to modelling multilevel selection (Kerr and Godfrey-Smith 2002), Maynard Smith demonstrates this admirably (Maynard Smith 2002). Discussing how relatedness can be modelled either using a gene-centred approach or through an individual-centred approach using inclusive fitness, he notes that in general having multiple mathematically equivalent models in mind at the same time can be very helpful in making inferences about what is going on in a given selective episode. He then remarks:

> I confess that in this case I find the gene-centred approach both mathematically simpler and causally more appropriate, but this may merely reflect the fact that I prefer microscopic to holistic models: Maxwell-Boltzmann to classical thermodynamics, and Dawkins to Price's equation.　　(Maynard Smith 2002).

Just because two abstract mathematical models yield the same prediction it does not mean they are causally equivalent. Regardless of your views on the matter, Maynard Smith's attitude is exactly the kind of honest philosophical transparency that we need more of in evolutionary theory.

3.5 Genetic determinism

One of the most thoughtful critics of the gene's-eye view over the years was Patrick Bateson. Bateson, whose grandfather's cousin William Bateson gave science the word 'genetics', was an English behavioural ecologist with a special interest in the developmental biology of animal behaviour. Ever since he wrote a critical review of *The Selfish Gene* (Bateson 1978), Bateson has been a reliable sparring partner of Dawkins. When he contributed to the *festschrift Richard Dawkins: How a Scientist Changed the Way We Think* (Grafen and Ridley 2006), he was grouped under heading 'Antiphonal voices' (Bateson 2006a). The main message of Bateson's critique was that too much focus on genes over other factors contributing to the formation of an animal obfuscates the true causes of development and so results in a poorer account of evolution.

One of Bateson's concerns was that the intentional language of the gene's-eye view may result in confusion regarding genetic determinism, the idea that genes and only genes matter in determining phenotypes. 'Dawkinsspeak leads to Dawkinspractice', as Gray (1992) once put it. It's not that gene-selectionists are committed to genetic determinism, Bateson argued, but that sloppy terms like 'gene for', 'blueprint', 'genetic program', and 'hardwiring' mean that one can be forgiven for thinking that they are (Bateson 1981, 1999, 2001; Curley et al. 2009; see also Noble 2015). It is worth remembering that these terms were, and continue to be, widely used also by critics of the gene's-eye view. For example, the 'genetic program' metaphor is thought to have been independently coined in two papers, one by François Jacob and Jacques Monod and one by Ernst Mayr (Peluffo 2015). It should also be noted that Bateson described Chapter 2 of *Extended Phenotype* as the best rebuttal of genetic determinism he had ever come across (Bateson 1986; even arch-critic Mary Midgley described it as 'admirable', Midgley 1983). Even so, Bateson argues that 'gene-for'-language is slippery. For example, Bateson (1986) criticizes Brian Charlesworth (1978) for referring to a gene 'coding' for altruism. Bateson contends that this implies that there is a one-to-one correspondence between gene and behaviour. Thus, even the most careful of population geneticists may appear as if they forgot that they are dealing with a shorthand.

In general, Bateson thought that gene selectionists' focus on how replicators use bodies to produce copies of themselves gives too much credit to genes in the developmental process. In a classic paragraph in his review of *The Selfish Gene* he wrote:

> A legitimate focus on gene's intentions should not be used as an excuse for resuscitating moribund preformationism. (...) Dawkins accepts all this but then reveals his uncertainty about which language he is using by immediately giving special status back to the gene as the programmer. Consider a case in which the ambient environmental temperature during development is crucial for the expression of a particular phenotype. If the temperature changes by a few degrees the survival machine is beaten by another one. Would not that give as much status to a necessary temperature value as to a necessary gene? The temperature value is also required for the expression of a particular phenotype. It is also stable (within limits) from one generation to the next. It may even be transmitted from one generation to the next if the survival

machine makes a nest for its offspring. Indeed, using Dawkins' own style of teleological argument one could claim that *the bird is the nest's way of making another nest.*   (Bateson 1978; my emphasis)

Dawkins first response to Bateson channelling his inner Samuel Butler was to argue that the nest analogy fails (Dawkins 1978). Nests are not replicators, as a mutation in the nest—'for example the accidental incorporation of a pine needle instead of the usual grass'—is not transmitted to offsprings' nests. In other words, whereas one can plausibly talk about a causal arrow between gene and bird, there is none in the reverse direction. While true, this was not really Bateson's point, which Dawkins would later acknowledge (Dawkins 1982a, p. 98).

Again, this is partly a matter of what aspect of the biology one is interested in. Whereas development requires biologists to pay equal attention to both genetic and environmental factors, proponents of the gene's-eye view have argued that the unit of selection question does not, as only genes are replicated and transmitted across generations. Dawkins makes this point with the following backhanded compliment:

> As is so often the case, an apparent disagreement turns out to be due to mutual misunderstanding. I thought Bateson was denying proper respect to the Immortal Replicator. Bateson thought that I was denying proper respect to the Great Nexus of complex causal factors interacting in development.   (Dawkins 1982a, p. 99)

The Bateson–Dawkins sparring would continue some three decades later. Discussing the revived interest in epigenetics, Dawkins summarized the focus on such developmental details as an 'obscurantist red herring' (Dawkins 2004b). He also retold a story of how his Oxford graduate students used to describe members of Bateson's Cambridge department as being adherents to the dictum: 'Never use a simple explanation if a more complicated one will do instead.' As Bateson admits, this dictum had self-mockingly been endorsed by the Sub-Department of Animal Behaviour at Cambridge in honour of its doyen Robert Hinde, who was known to repeatedly emphasize that

'Behaviour is complicated' (Bateson 2006a, 2006b). In return, Bateson offered The Oxford Principle: 'Never use a causal explanation if a teleological one will do instead' (Bateson 2006a). Bateson's point was that while the intentional language may be a powerful heuristic, it is much easier to come up with explanatory stories, which may or may not be true, than Dawkins acknowledges. Behaviour is complicated and a one-sided focus on genes misses this.

3.5.1 Evo-devo and developmental systems theory

Whether a focus on genes as the unique bearers of heredity is still tenable is a hotly contested question in contemporary evolutionary biology. Bateson is far from the only one who has argued that the gene's-eye view, and indeed the broader field of evolutionary biology since the modern synthesis, has paid too little attention to developmental biology and related concepts (see, e.g., Jablonka and Lamb 2005 and Bonduriansky and Day 2018).

The developmental charge has been organized under two main labels: evolutionary developmental biology (evo-devo) and developmental systems theory. While proponents of both evo-devo (Müller 2007; Carroll 2008) and developmental systems theory (Oyama 1985; Ford and Lerner 1992; Gray 1994, 2001; Griffiths and Gray 1994) have criticized gene-centred approaches to biology using a combination of empirical and conceptual claims, the latter have done so in a more radical way. To what extent this is a consequence of evo-devo growing out of biology and developmental systems theory out of philosophy, as suggested by Robert et al. (2001), I am not sure. Nevertheless, perhaps because of its more radical nature, developmental systems theory has engaged more directly with the conceptual assumptions and implications of the gene's-eye view.

The central claim of developmental systems theory is that genes are not causally special; they do not have a privileged role during development. The notion at the heart of this critique is that offspring inherit more than genes from their parents, including but not limited to methylation patterns, gut microbes, other contributions from maternal diet, and ecological habitats, none of which can be said to be more important than others. Organisms are no more vehicles for replicators than they are for language or other things. This is sometimes known as the 'parity argument', following Susan Oyama's very influential book *The Ontogeny of Information* (Oyama 1985). By this argument, the fundamental unit of evolution is not the gene but the life cycle, and the developmental system is the matrix containing all the resources needed to reproduce the life cycle.

By the term ontogeny of information, Oyama takes aim at the gene's-eye view gene as a package of information, not a scrap of nucleic acid. As Williams noted: 'The gene is a package of information, not an object. The pattern of base pairs in a DNA molecule specifies the gene. But the DNA molecule is the medium, it's not the message' (Williams 1996b). According to Williams, what is passed on from one generation to the next is information for how the organism should develop. In contrast, Oyama posits that this information is reconstructed during development every generation. To make sense of Oyama's point, Sterelny and Griffiths suggested an analogy of a rat memorizing how to solve a maze (Sterelny and Griffiths 1999, p. 101). Instead of having the correct route carefully marked out on a map in their brain, the rat relies on cues from the external environment, such as the structure of the maze, combined with memories from previous attempts of solving the maze, to reconstruct the route. The key is that the route information is reconstructed every time, not passed on intact. In the same way, Oyama argued, embryonic development does not rely on a filled out developmental programme—there is no preformed plan—but genes interact with other informational sources within and outside the embryo itself.

Developmental system theorists also reject the standard nature versus nurture dichotomy. Their argument is not just that most traits are the product of both, as well as their interaction. That, no one disputes. Rather, they argue that it leads to the wrong questions being asked (Griffiths and Taberny 2013). In Chapter 1, I argued that Fisher's paper

'The correlation between relatives on the supposition of Mendelian inheritance' (Fisher 1918) marked the beginning of the gene's-eye view, partly thanks to his introduction of the concept of variance. Fisher and subsequent researchers used this idea to ask how much difference in a trait among individuals in a population was attributable to genetic differences and how much was due to environmental differences. To developmental system theorists, this dichotomous view of nature and nurture results in an unproductive focus on *how much* rather than *how* development unfolds.

The interest that developmental systems theorists have shown in the gene's-eye view appears not to have been reciprocated by gene selectionists. One exception is Sterelny et al. (1996), who argued that by extending the replicator concept, one could incorporate most insights from developmental systems theory within the replicator-vehicle framework (see Lu and Bourrat 2018 for a similar argument and Griffiths and Gray 1997 for a reply). While it is important to explore how different frameworks can be integrated, it is not always crucial that they are. Just as I agree with Maynard Smith's discussion of different ways of modelling social evolution outlined above, I share much of Haig's assessment that developmental systems theory is logically coherent and likely a more suitable heuristic framework for certain questions in biology than the genic perspective (Haig 2012).

3.6 Human nature and human affairs

In a letter to his good friend, the botanist Joseph Dalton Hooker, written 15 years before the publication of *The Origin of Species*, Darwin described his feelings concerning his theory of evolution like 'confessing a murder' (Darwin Correspondence Project 2020c). The implications of Darwin's theory for science, philosophy, and religion were huge (Ruse 1986; Dennett 1995; Sober 2011). As Darwin jotted down in a notebook in 1838: 'Origin of man now proved. Metaphysic must flourish. He who understands baboon would do more towards metaphysics than Locke' (Darwin 1838).

Evolutionary biology has had a rocky relationship with the concept of human nature and human affairs ever since. On page 1 of *The Selfish Gene,* Dawkins approvingly quotes George Gaylord Simpson's comment that all attempts to answer the question 'what is man?' prior to 1859 were 'useless' and 'best ignored'. The anthropologist Marshall Sahlins could not have disagreed more. In his *The Use and Abuse of Biology: An Anthropological Critique of Sociobiology,* he wrote of evolutionary approaches to human behaviour as contributing 'primarily to the final translation of natural selection into social exploitation' (Sahlins 1976, p. 73).

Central to these disparate opinions was the issue of genetic determinism. As Lewontin, Rose, and Kamin put it in *Not in Our Genes*:

> Sociobiology is a reductionist, biological determinist explanation of human existence. (...) The academic and popular appeal of sociobiology flows directly from its simple reductionist program and its claim that human society as we know it is both inevitable and the result of an adaptive process. (Lewontin, Rose and Kamin 1984, p. 236).

A few years earlier, in the winter of 1981, Rose had written to *Nature* to alert readers of an article in *New Nation*, a magazine associated with the UK far-right organization National Front, titled 'Nationalism, racialism: products of our selfish genes' (Rose 1981). The article included a photo of Dawkins and cited the works of Dawkins, E.O. Wilson, Maynard Smith, and one 'Travers' (which, Rose points out, presumably was meant to be Trivers) as bolstering their white supremacy position. Rose's letter ended with the following appeal: 'May I suggest that it would be in the public interest that John Maynard Smith and Richard Dawkins should clearly dissociate themselves from the use of their names in support of this neo-Nazi balderdash' (Rose 1981). Dawkins (1981b) and Maynard Smith (1981) both promptly, and without hesitation, did just that, saying that 'there is nothing in modern evolutionary biology which leads to that conclusion [that our genetic constitution makes it impossible for us to live in a racially integrated society]' (Maynard Smith 1981). This was not the first time that Rose had made Maynard Smith publicly distance himself from the misrepresentations of his work. A year and a half earlier, in June 1979, Maynard Smith had typed a letter to *New Scientist* that the editors gave the title 'Thatcher's biology':

> When, in 1964, I wrote a largely mathematical paper in *Nature* on 'Group and kin selection' I did not know that I was starting a chain of events that would lead to the election of Mrs Thatcher's government. Indeed, I only realised it when I read Steven Rose's 'The Thatcher view of human nature' (Forum, 18 May, p 575). I yield to no one, not even to Rose, in my dislike of Mrs Thatcher's declared policies, but what would he have me do? Fiddle the algebra?   (Maynard Smith 1979)

As discussed earlier, the charge of genetic determinism is largely unfair. While proponents of the gene's-eye view may assign too much causal power to genes, they are not denying a role of the environment. Genes are difference makers, but that does not mean that only genes matter (as Bateson noted, any doubters should consult chapter 2 in *The Extended Phenotype*). Yet, several other books on this theme were published around the same time. Examples included *Against Biological Determinism* (Rose 1982a), *Towards a Liberatory Biology* (Rose 1982b), and *The Dialectical Biologist* (Levins and Lewontin 1985). However, it was *Not in Our Genes* that made Dawkins 'stand up and be counted' as a sociobiologist, something he had resisted due to his dislike of the term (Dawkins 1985; see also Dawkins 1986b). The many ins and outs of this historical period are covered in an excellent, comprehensive way in Ullica Segerstrale's *Defenders of the Truth: The Sociobiology Debate* (Segerstrale 2000). Here, I will focus on one specific implication of evolutionary biology in general and the gene's-eye view in particular: is the universe naturally good or bad or indifferent?

3.6.1 T.H. Huxley and Darwinian nightmares

Upon reading *The Selfish Gene*, Randolph Nesse—one of the founders of the field of evolutionary medicine and co-author with George Williams of *Why We Get Sick: The New Science of Darwinian Medicine*

(Nesse and Williams 1994)—described how it left him with weeks of dreaming about lumbering robots (Nesse 2006). He was not alone. In the preface of *Unweaving the Rainbow*, Dawkins describes how a foreign publisher could not sleep for three days after reading *The Selfish Gene*, he was so disturbed by what he considered to be the cold and bleak implications of the book (Dawkins 1998b, p. ix). What were these implications?

The Selfish Gene is not especially concerned with humans, a point lost on many commentators on moral matters (but see Williams 1976 for an exception). On this issue Dawkins clearly differs from E.O. Wilson, who followed *Sociobiology* with *On Human Nature*, a book that explicitly deals with how sociobiology applies to humans (Wilson 1978). Instead, the bleakness that so many readers found so repugnant came from a more general view of the universe. In *River Out of Eden* Dawkins writes that 'a universe of electrons and selfish genes, blind physical forces and genetic replication (...) has precisely the properties we should expect if there is, at bottom, no design, no purpose, no evil, no good, nothing but pitiless indifference' (Dawkins 1995b, p. 133). Similarly, in *A Devil's Chaplain* he writes that 'Blindness to suffering is an inherent consequence of natural selection. Nature is neither kind nor cruel but indifferent' (Dawkins 2003, p. 9). The title itself comes from another letter from Darwin to Hooker, in which he wrote: 'What a book a devil's chaplain might write on the clumsy, wasteful, blundering low and horridly cruel works of nature!' (Darwin Correspondence Project 2020d).

Williams's views were along the same lines. He described mother nature as a 'wicked old witch' and considered natural selection to be 'a morally unacceptable process' (Williams 1993) that could 'honestly be described as a process for maximizing short-sighted selfishness' (Williams 1989). For this reason, both Williams and Dawkins were also highly critical of the Gaia hypothesis (Lovelock and Margulis 1974; Lovelock 1979). The Gaia hypothesis is the idea that Earth and its living organisms interact in a synergistic self-regulating way in order to maintain life and is sometimes seen as alternative conception of the world, a more romantic one, focused less on conflict and competition and more on cooperation and harmony. Williams and Dawkins forcefully argued that it contradicted evolutionary theory (e.g. Dawkins 1982a, pp. 234–237; Williams 1993).

Dawkins and Williams were not the first to make the case that the world is a ruthless place. The phrase 'nature red in tooth and claw' appeared in Lord Tennyson's poem 'In Memoriam A.H.H.' in 1850, a message Dawkins (1976, p. 2) and Williams (1989) both endorsed as providing an accurate summary of the modern understanding of natural selection. More than Lord Tennyson, however, T.H. Huxley was the premier spokesperson for this world view. Mary Midgley even went so far as to call Dawkins a 'Huxleyean rather than a Darwinian thinker' (Midgley 2010, p. 47). Given that Dawkins is often accused of being a naive adaptationist, the moniker is a little strange. As noted in Chapter 1, Huxley had serious doubts about the power of natural selection and was not particularly impressed by adaptations. Furthermore, the influence of Huxley was probably even greater on Williams. After all, he co-edited a new edition on Huxley's celebrated Romanes Lectures, *Evolution and Ethics*, delivered in Oxford in 1892 and published 2 years later (Paradis and Williams 1989). While Huxley and contemporary Social Darwinists like Herbert Spencer shared a generally Hobbesian view of nature as being 'on about the same level as a gladiator show' (Huxley 1888, p. 200), he firmly rejected the implication that this is how it therefore should be. Instead, as he argued in his Romanes lecture, the natural process needs to be defeated: 'Let us understand, once for all, that the ethical progress of society depends, not on imitating the cosmic process, still less in running away from it, but in combating it.' (Huxley 1894, p. 83). Williams strongly approved of Huxley's message:

> With what other than condemnation is a person with any moral sense supposed to respond to a system in which the ultimate purpose in life is to be better than your neighbor at getting genes into future generations, in which those successful genes provide the message that instructs the development of the next generation, in which that message is

always 'exploit your environment, including your friends and relatives, so as to maximize our genes' success, in which the closest thing to a golden rule is 'don't cheat, unless it is likely to provide a net benefit'?

(Williams 1997, p. 154)

Someone who did not care much for Huxley's message was the Russian aristocrat, anarchist, and scientist Peter Kropotkin. Outraged by Huxley's 'The struggle for existence in human society' quoted above (Huxley 1888), Kropotkin developed his argument in eight essays during the 1890s, eventually jointly published in *Mutual Aid: A Factor of Evolution* (Kropotkin 1902). Relying primarily on his own fieldwork in the remote parts of Siberia, Kropotkin's central thesis was that relationships among organisms were best characterized not by a struggle for existence, but by mutual solidarity. Making sense of Huxley and Kropotkin and the broader connection between evolutionary theory and ethics is far from trivial (see extensive treatments by, e.g., Rachels 1991; Nitecki and Nitecki 1993; James 2010).

Like Kropotkin, several modern writers have also sought to state the case that the world is not fundamentally selfish. Books making this argument have often used variations on the title of *The Selfish Gene* to make the point. Examples include Joachim Bauer's *Das kooperative Gen: Abschied vom Darwinismus* (The Cooperative Gene: Farewell to Darwinism; Bauer 2008), Joan Roughgarden's *The Genial Gene* (Roughgarden 2009), Fern Elsdon-Baker's *The Selfish Genius* (Elsdon-Baker 2009), and Göran Greider's *Den solidariska genen* (The Loyal Gene; Greider 2014). In fairness, books supportive of a gene's-eye view have had similar titles. One example is Dawkins's former student, Mark Ridley's *Mendel's Demon: Gene Justice and the Complexity of Life* (Ridley 2000), which was renamed *The Cooperative Gene: How Mendel's Demon Explains the Evolution of Complex Beings* for the American market.

Others have also worried about the association between selfish genes and selfish humans. Andrew Briggs, Hans Halvorson, and Andrew Steane suggest replacing 'selfish genes' with 'eager genes' simply because the latter is 'morally neutral' (Briggs et al. 2018, pp. 201–202). Another example is the group selection advocate David Sloan Wilson, who has argued that the gene's-eye view (as well as inclusive fitness and game theory) is simply a way to transform altruism into selfishness (Wilson 2015a, p. 32). Finally, Roughgarden (no friend of group selection) has described orthodox sexual selection theory, which she places in the same intellectual camp as the gene's-eye view under the label 'neo-Spencerism' (from Herbert Spencer), as being underwritten by 'genetic classicism' and 'genetic entitlement' (Roughgarden 2009, p. 4). Further, she argues that 'if sexual selection is indeed true, then so be it; and the prospect of an egalitarian society is an unrealistic mirage' (Roughgarden 2009, p. 5).

To me, there is no better response to these worries than the one developed by David Hume and G.E. Moore. In a short paragraph in *A Treatise of Human Nature*, Hume argued that you cannot coherently move from a statement about how the world *is* to a statement about how the world *ought* to be (Hume 1793, Book III, Part I, Section I). A century later, Moore developed a similar idea in *Principia Ethica* and coined the term naturalistic fallacy to refer to the mistake (Moore 1903). Dawkins was very much on the same page as Huxley and Williams, as well as Hume and Moore. Discussing the prospect of going from evolution to ethics he stated:

> As an academic scientist I am a passionate Darwinian, believing that natural selection is, if not the only driving force in evolution, certainly the only known force capable of producing the illusion of purpose which so strikes all who contemplate nature. But at the same time as I support Darwinism as a scientist, I am a passionate anti-Darwinian when it comes to politics and how we should conduct our human affairs.'
> (Dawkins 2003, pp. 10–11)

Indeed, Dawkins ends the first edition of *The Selfish Gene* with:

> We have the power to defy the selfish genes of our birth and, if necessary, the selfish memes of our indoctrination. (...) We are built as gene machines and cultured as meme machines, but we have the power to turn against our creators. We, alone on earth, can rebel against the tyranny of the selfish replicators.   (Dawkins 1976, p. 215)

The way we talk about evolution has ramifications for how we view ourselves. Yet, just as there is no contradiction in the physician scientist devoting her career to both describing and eradicating cancer, our commitment to the science of evolution says nothing about our moral outlook. Selfish genes do not necessarily make selfish people.

3.7 Summary

- Anthropomorphizing is a much-maligned concept in biology, but when paired with mathematical modelling (licensed anthropomorphizing) it can be a very powerful thinking tool.
- While the gene's-eye view is most straightforward under additivity, complex genetic interactions, such as epistasis, dominance, or heterozygote advantage, are not as big a problem as some critics make them out to be.
- The bookkeeping objection is one of the most famous criticism of the gene's-eye view. It holds that while all evolutionary change can be recorded as changes in allele frequencies, it is incorrect to therefore assign causality to the level of genes. Disagreements over the merit of this objection to the gene's-eye view led to a fertile philosophical debate over pluralism in evolutionary explanations.
- Developmental biologists have long been critical of what they perceive as the gene's-eye view's overly narrow focus on genes' causal primacy in development. The most ambitious form of this critique has been organized as developmental systems theory, which offers a radically different approach to biology that also has heuristic value.
- The gene's-eye view was also entangled in debates over human nature. Both Williams and Dawkins express a rather bleak Huxleyan view on the universe and agreed that it is up to us to be the good we want to see in the world.

4

Inclusive Fitness and Hamilton's Rule

4.1 Introduction

To explain what was so special about her mentor W.D. (Bill) Hamilton, Marlene Zuk wrote that the difference between him and everyone else was 'not the quality of his ideas, but their sheer abundance' (Zuk 2000). The proportion of his ideas that were actually any good was about the same as anyone else, instead:

> the difference between Bill and most other people was that he had a total of over one hundred ideas, with the result that at least ten of them were brilliant, whereas the rest of us have only four or five ideas as long as we live, with the result that none of them are.   (Zuk 2000)

Hamilton indeed had number of brilliant ideas. He made substantial contributions to the study of the origin of sex, sex ratios, genomic conflicts, host–parasite interactions, and the evolution of senescence (Grafen 2004; Segerstrale 2013). His most influential idea, and the one that bears his name, is about the evolution of social behaviour, especially altruism and showed that it is possible to evolve social behaviour that comes with a fitness cost if the beneficiary of that behaviour is a relative (Hamilton's Rule).

Many are the biology students who first learned about the intricacies of Hamilton's work through the writings of Dawkins. The first

The Gene's-Eye View of Evolution. J. Arvid Ågren, Oxford University Press. © J. Arvid Ågren 2021.
DOI: 10.1093/oso/9780198862260.003.0005

draft of *The Selfish Gene* was based on a series of lecture notes Dawkins developed to communicate Hamilton's ideas to his undergraduate students (Dawkins 2013a, p. 201). One of these former students, Jim Mallett, now a professor at Harvard University and University College London, recalls the name of the course to have been 'Gene Machines' (J. Mallet, personal communication). The course title later inspired the name of Chapter 4 of *The Selfish Gene* (The gene machine).

If Dawkins clearly cherished teaching, especially within the Oxford tutorial system (Dawkins 2008a), Hamilton had a different reputation. He was a notoriously bad lecturer, at least as far as large undergraduate courses went. During his time as a junior faculty member at Imperial College, Silwood Park, undergraduates used to urge the departmental administrators not to count their results in Hamilton's course towards their degree marks (Segerstrale 2013, p. 392). Robert Trivers also recalls Hamilton coming to give a talk at Harvard only to deliver 'one of the worst lectures [Trivers] had ever seen in [his] life' (Trivers 2015, p. 188).

Many biologists therefore seem to have learned about Hamilton's work from secondary sources. The fact that Hamilton's two-part landmark paper from 1964 (Hamilton 1964a, 1964b) has often been cited incorrectly has been taken as evidence of this. The correct title is 'The genetical evolution of social behaviour' but, as Seger and Harvey (1980) showed, out of 200 papers published between 1965 and 1979 some 20% cited it as 'The genetical theory of social behaviour' instead. The paper was miscited by all sorts of authors, critics and supporters alike, and among graduate students as well as luminaries including Stephen Jay Gould, Richard Lewontin, E.O. Wilson, and John Maynard Smith. Seger and Harvey noted that among the sixty-three papers published before 1975 only one contained the incorrect citation, leading them to suggest that many had used the erroneous version found in the first edition of E.O. Wilson's *Sociobiology* (Wilson 1975). The first edition of *The Selfish Gene* also contains this mistake, and whether the error came from either book or not is unclear. In the endnotes added to the second edition of *The Selfish Gene* published in 1989, Dawkins devotes quite a bit of space to rebutting what he considers to be a 'memetic myth' (Dawkins 1989, p. 326). The mutant title, Dawkins argues, is just one word off and its similarity to Fisher's *The Genetical Theory of Natural Selection* makes multiple independent errors quite likely.

Dawkins was also instrumental in rectifying public misunderstandings of Hamilton's work. One good example is his 'Twelve misunderstandings of kin selection' (Dawkins 1979). In it, he outlined and refuted common mistakes other researchers had made in print. For example, misunderstanding number 3 was 'The theory of kin selection demands formidable feats of cognitive reasoning by animals'. An example of this came from the earlier mentioned *The Use and Abuse of Biology*, where Marshall Sahlins wrote of the 'considerable mysticism' involved in Hamilton's work:

> In passing it needs to be remarked that the epistemological problems presented by a lack of linguistic support for calculating, r, coefficients of relationship, amount to a serious defect in the theory of kin selection. Fractions are of very rare occurrence in the world's languages, appearing in Indo-European and in the archaic civilizations of the Near and Far East, but they are generally lacking among the so-called primitive peoples. Hunters and gatherers generally do not have counting systems beyond one, two and three. I refrain from comment on the even greater problem of how animals are supposed to figure out how that r (ego, first cousins) $= 1/8$.   (Sahlins 1977, pp. 44–45).

In response, Dawkins wryly asked where, given the logarithmic spiral of its shell, a snail was meant to keep its log tables? Some of the listed misunderstandings may seem a little silly when read today. Nevertheless, they do persist. Take, for example, misunderstanding number 5: 'All members of a species share more than 99% of their genes, so why shouldn't selection favour universal altruism?' This seems to have to have originated with Washburn (1978) but was perpetuated as late as 2004 by Niles Eldredge (in a book with the subtitle 'Rethinking sex and selfish genes', to boot; Eldredge 2004, pp. 42–43).

There is thus a long and intimate association between the gene's-eye view and the concept of inclusive fitness and Hamilton's Rule. In this chapter, I examine that relationship. I first introduce what I call Hamilton's unfinished revolution. Hamilton provided one of the first clear articulations of the gene's-eye view but, much to the frustrations of the likes of Dawkins and Maynard Smith, never completely switched to a fully gene-centric approach. More generally, proponents of the gene's-eye view and inclusive fitness have emphasized their equivalence, while also stressing the limitations of the other. The connection between the two is partly a product of personal connections between key individuals involved, but there is also a more formal link. To demonstrate this, I outline a derivation of the so-called general version of Hamilton's Rule, which looks rather different from the informal version ($rb > c$) that makes an appearance in most introductory evolution courses. This general version is also at the centre of much of the current kerfuffle over Hamilton's work and I explain why this is and how the gene's-eye view fits into those debates.

Central to many disagreements over inclusive fitness is the issue of how to scientifically handle the neo-Paleyan fascination with the appearance of design that is such a key motivator for both the gene's-eye view and inclusive fitness theory. One attempt is Alan Grafen's ambitious Formal Darwinism Project, which seeks to construct a mathematical bridge between the dynamics of population genetics and the appearance of design. I discuss how this project provides an interesting window into the relationship between the gene's-eye view and inclusive fitness. In particular, Grafen's solution—that organisms should behave as if designed to maximize their inclusive fitness—highlights the need for unity of purpose among genes. How to handle what happens when this unity breaks down (when genes are in conflict) and strikes at core of the connection between genic and the organismic approaches.

Examinations of the relationships between the gene's-eye view and inclusive fitness continue to this day. I end the chapter by discussing two recent attempts at reconciling insights from the two perspectives and what light these arguments can shed on the gene's-eye view.

4.2 The origin and diversity of Hamilton's Rule

Hamilton's life in biology began at the University of Cambridge. While there, he discovered R.A. Fisher; first by finding *The Genetical Theory of Natural Selection* in the library and later the man himself, then a retired professor in the Genetics Department. The effect that *The Genetical Theory* had on the young Hamilton is almost impossible to exaggerate. One need only read Hamilton's endorsement on the back of the 1999 *Complete Variorum Edition* of the book:

> This is a book which, as a student, I weighed as of equal importance to the entire rest of my undergraduate Cambridge BA course and, through the time I spent on it, I think it notched down my degree. Most chapters took me weeks, some months; even Kafka whom I read at the same time couldn't depress me like Fisher could on say, the subject of charity, nor excite me like his theory of civilization. Terrify was even the word in some topics and it still is, so deep has been the change from all I was thinking before. (…) By the time of my ultimate graduation, will I have understood all that is true in this book and will I get a First? [the highest level of degree a UK undergraduate can receive] I doubt it. In some ways some of us have overtaken Fisher; in many, however, this brilliant, daring man is still far in front.   (Hamilton 1999)

Reading Fisher not only cemented Hamilton's interest in evolution but also planted the seed of what would be his obsession for the early years of his career: the genetics of altruism. After Cambridge, he moved to London for his graduate research, but his time in the capital appears to have been miserable. He was often so lonely that he lingered in underground stations and public buses in order to receive some human interactions (Segerstrale 2013, p. 64). Still, by any standard, his PhD work has had far-reaching influence. On 7th March 1963, he wrote to Colin Hudson, his close friend from their Cambridge undergraduate days, telling him that: 'My "letter" has been accepted by the "American Naturalist", which is very encouraging (…) I am still working on the last section of my main paper—the real manifesto' (Reprinted in Segerstrale 2013, p. 85). The 'letter' would appear as 'The evolution of altruistic behavior' later the same year. Towards the end of his career, Hamilton would refer to the paper as his 'little-read first paper' (Hamilton 1996, p. 5). He had actually completed 'the real manifesto' prior to penning his letter, but a reviewer (who we now know was John Maynard Smith) had suggested extensive rewrites, including splitting the paper into two (hence, Hamilton 1964a, 1964b). Feeling the pressure from his graduate advisors, he decided to submit a short précis of his key ideas. After a rapid desk rejection from *Nature*—which was accompanied with the advice to try a 'psychological or socialogical [sic] journal' (Hamilton 1996, p. 3)—he decided to send the 'letter' to the *American Naturalist*.

Despite being less than three pages long, the paper is remarkably full of novel insights (Gardner 2015; Marshall 2015, pp. 11–13). For instance, the paper contains one of the clearest and earliest articulations of the gene's-eye view. Discussing a locus where the G allele causes its carrier to perform some sort of altruistic behaviour, whereas *g* has no such effect, he writes:

> Despite the principle of 'survival of the fittest' the ultimate criterion that determines whether G will spread is not whether the behavior is to the benefit of the behaver but whether it is of benefit to the gene G (…) With altruism this will happen only if the affected individual is a relative of the altruist, therefore having an increased chance of carrying the gene, and if the advantage conferred is large enough compared to the disadvantage to offset the regression, or 'dilution,' of the altruist's genotype in the relative in question.   (Hamilton 1963)

In only a few sentences, Hamilton presents the profound insight that from a gene's point of view it does not matter if you are transmitted through the body in which you reside, or through a copy of yourself that is present in another individual. To capture this insight quantitatively, he introduced the inequality

$$k > 1 / r$$

where k is the ratio of fitness benefit to recipient and fitness cost to actor and r the degree of relatedness. This simple expression describes the necessary and sufficient conditions for a gene to be favoured by natural selection. Now known as Hamilton's Rule, a term first coined by Charnov (1977), it is usually expressed in the simpler form $rb > c$, where r is still the degree of relatedness, and b is the benefit to the recipient and c is the cost to the actor.

Hamilton considered his work an 'extension of the classical theory' of natural selection (a coy reference to the poor reputation of Fisher's Fundamental Theorem; Hamilton 1963). In addition to Hamilton's Rule, the key part of this extension was the concept of inclusive fitness. The following year, in 1964, in the majestic two-part paper for *The Journal of Theoretical Biology,* Hamilton defined it thus:

> Inclusive fitness may be imagined as the personal fitness which an individual actually expresses in its production of adult offspring as it becomes after it has been first stripped and then augmented in a certain way. It is stripped of all components which can be considered as due to the individual's social environment, leaving the fitness which he would express if not exposed to any of the harms or benefits of that environment. This quantity is then augmented by certain fractions of the quantities of harm and benefit which the individual himself causes to the fitnesses of his neighbors. The fractions in question are simply the coefficients of relationship appropriate to the neighbours whom he affects.   (Hamilton 1964a)

More informally it can be summarized as:

$$\text{Inclusive fitness} = \text{direct fitness} + \text{indirect fitness}$$

To calculate an individual's inclusive fitness you first determine its direct fitness, to which you add those parts of the fitnesses of others for which the focal individual is causally responsible (scaled by relatedness; the indirect fitness). Finally, you then subtract ('strip') the part of the personal fitness that is due to the social environment of the focal individual.

The last stripping step is to avoid the same fitness being counted twice, both as the focal individual's and that of a social partner, as if a given offspring had multiple existences (so-called double counting; Grafen 1982; Queller 1996; see also Levin et al 2019 for a recent example of how to handle this issue in practice). Because the pheno-types of relatives are correlated due their shared ancestry, this subtraction is essential in order to properly separate causal from correlational effects. By ignoring fitness effects due to actions beyond the focal individual, it also keeps inclusive fitness under the full control of the individual.

The central message here is that an individual organism may caus-ally affect the transmission of a given gene both through having off-spring on its own and by helping relatives increase their reproductive success. As shown by Hamilton, depending on preferences, this pro-cess can be captured either from the perspective of an individual organism by using inclusive fitness or from a gene's-eye view. Crucially, Hamilton considered inclusive fitness to be a causal property of the individual organism, a fact that will become relevant later on.

In addition to Hamilton's Rule and inclusive fitness, the third term to come out of Hamilton's work on social evolution is kin selection. Not a term coined by Hamilton himself, but by Maynard Smith (1964), kin selection is selection involving interactions among genea-logical kin. Kin selection has been less controversial than its two sister concepts, likely a consequence of it being an empirical process, rather than a mathematical framework (Birch 2017a). As will become clear, the mathematical properties of Hamilton's Rule and inclusive fitness leave much room for disagreements.

4.2.1 Hamilton's Rule today

Hamilton's Rule and inclusive fitness were criticized from the get-go. Theoretical population geneticists, who saw little value in the concept compared to more exact modelling approaches were especially crit-ical (examples included Cavalli-Sforza and Feldman 1978; Feldman and Cavalli-Sforza 1981; Karlin and Matessi 1983). The beginning of the most recent iteration of the debate is usually dated to the publica-tion of a paper titled 'The evolution of eusociality' in *Nature*, by E.O. Wilson and two mathematicians turned mathematical biologists,

Corina Tarnita and Martin Nowak (Nowak et al. 2010). The paper led to a flurry of back-and-forth argument and counter-arguments (see, e.g., Abbott et al. 2011; Gardner et al. 2011; Allen et al. 2013; Marshall 2015; Liao et al. 2015; Nowak and Allen 2015; van Veelen et al. 2017 for some of the original arguments; and Birch 2014, 2017a, 2017b, Birch and Okasha 2015 for overviews).

The current controversy can be illustrated by contrasting statements from leading authors on each side of the debate. On the one hand, in the paper that kicked off the recent spat, Nowak et al. state that Hamilton's Rule 'almost never holds'. On the other hand, 137 authors described Nowak et al.'s statement as 'based on a misunderstanding of evolutionary theory and a misrepresentation of the empirical literature' (Abbot et al. 2011). Similarly, Gardner et al. (2011) characterized it as 'simply incorrect' and go on to argue that Hamilton's Rule has 'the same generality and explanatory power as the theory of natural selection itself'. To see why the tone has become so derisive, it is instructive to go through the assumptions of the contested version of Hamilton's Rule, the so-called general version. These assumptions are what make the version so attractive for its supporters and so unbearable for its critics (Box 4.1).

The criticism stemming from Nowak et al. (2010) of the general version of Hamilton's Rule stands on three legs (Birch 2017b, pp. 64–76): it is tautological, it makes no predictions, and it fails to generate causal explanations. Crucial to all arguments are the minimal assumptions involved in deriving it. In fact, in one articulation of this criticism, Allen et al. (2013) go so far as claiming that it requires *no* assumptions. This is not quite true. Take, for example, Queller's decision to drop $E(w_i \Delta p_i)$ (Box 4.1). Because of phenomena such as selfish genetic elements, genetic transmission will be biased more often than many biologists probably appreciate, and $E(w_i \Delta p_i)$ will thus not equal 0. At the same time, the importance of selfish genetic elements is hardly at the heart of the disagreement between proponents and critics of Hamilton's Rule, so the relevance of this particular fact is somewhat limited. Nevertheless, a model is only ever as good as its

Box 4.1 The general version of Hamilton's Rule

The version of Hamilton's Rule at the heart of the disagreement takes a little getting used to. This way to capture Hamilton's Rule, sometimes referred to as the 'general model' as it is more general than Hamilton's own derivation, was developed by David Queller in the early 1990s (Queller 1992). Queller's paper is readable also for those of us without formal training in mathematical theory, but there are also several good primers on the model (see, e.g., Gardner et al. 2011 and Marshall 2015). Here, I follow the lead of Jonathan Birch, whose book *The Philosophy of Social Evolution* is a thorough introduction to the many conceptual issues underlying disagreements in social evolution research, especially those stemming from Hamilton's work (Birch 2017b).

Queller's derivation of Hamilton's Rule takes its starting point in the Price equation. The Price equation is an abstract statement about evolutionary change from one generation to the next (Price 1970; Frank 2012; Gardner 2020), such that:

$$\Delta \bar{p} = \frac{1}{\bar{w}}[Cov(w_i, p_i) + E(w_i \Delta p_i)] \tag{1}$$

The change in population frequency of a given allele ($\Delta \bar{p}$), where p_i is the individual gene frequency of the ith individual for the allele under consideration, is the sum of two population statistics. The first is the selection term, given by $Cov(w_i, p_i)$, the covariance between individual fitness w_i and individual gene frequency p_i. The second is the transmission term, $E(w_i \Delta p_i)$, the expected change in p_i between parent and offspring. Both terms are weighed by the population mean fitness $\bar{w}$.

With this in place, Queller makes three key assumptions when deriving Hamilton's Rule. First, p_i is reinterpreted as a breeding value, as used in quantitative genetics. This is possible because p_i can be thought of as not just an individual gene frequency, but as a linear combination of frequencies across multiple alleles of multiple loci. From this, $\Delta \bar{p}$ becomes the change in a quantitative polygenic trait.

(Continued)

Box 4.1. (Continued)

Next, the transmission term $E(w_i \Delta p_i)$ is assumed to be 0. This assumption has two key biological implications. First is that processes like segregation distortion, gametic selection, and genetic drift are assumed to be negligible and so safe to ignore. Second is that the average effects of alleles on the considered phenotype are constant. Because dominance and epistasis can both cause this assumption to be violated, the way to make sense of this assumption is to return to Fisher's expanded concept of the environment that was introduced in Chapter 1, which includes the rest of the genome. Both assumptions are quite substantial and strike at the heart of the relationship between Hamilton's Rule and the gene's-eye view. I will return to them later in the chapter.

Finally, with $E(w_i \Delta p_i)$ dropped, Queller takes $Cov(w_i, p_i)$ to be the effect of natural selection on the evolutionary change of the trait under study. This leaves:

$$\Delta \bar{p} = \frac{1}{\bar{w}}[Cov(w_i, p_i)] \tag{2}$$

To get from this to the familiar $rb > c$, the selective covariance term must be partitioned into rb and c. Queller achieves this by making use of the Lande–Arnold regression model of fitness (Lande and Arnold 1983). The fitness of the ith individual can therefore be stated using a linear regression model, so that:

$$w_i = \alpha + \beta_1 p_i + \beta_2 \hat{p}_i + \varepsilon_{wi} \tag{3}$$

Such linear function considers the gene frequency of a given individual (p_i), as well as the average individual's gene frequency of its social partners $(\hat{p}_i)$. It then captures the partial regression of an individual's fitness on that individual's gene frequency (β_i), accounting for that of the social partners and the partial regression of an individual's fitness on the gene frequency of social partners, this time accounting for that individual's gene frequency. α is the non-social part of fitness and taken to be the same for all individuals. ε_{wi} is the traditional error term of linear regressions and here it represents the discrepancy between actual and predicted fitness of the ith individual. This regression model is then substituted into the Price equation (2), leading to:

$$\bar{w}\Delta\bar{p} = Cov(\alpha, p_i) + \beta_1 Var(p_i) + \beta_2 Cov(\hat{p}_i, p_i) + Cov(\varepsilon_{wi}, p_i) \quad (4)$$

Simplifying and rearranging to state the condition for the population mean of the trait of interest to increase $\Delta\bar{p} > 0$ gives:

$$\Delta\hat{p} \succ 0 \Leftrightarrow \frac{Cov(\hat{p}_i, p_i)}{Var(p_i)}\beta_2 \succ -\beta_1 \quad (5)$$

Then, because $r, b, c,$ in Hamilton's Rule can be defined as:

$$r = \frac{Cov(\hat{p}_i, p_i)}{Var(p_i)}, \; b = \beta_2, \text{ and } c = -\beta_1$$

this notation means that (5) can be rewritten as:

$$(\Delta\bar{p}) \succ 0 \Leftrightarrow rb \succ c$$

That is, the frequency of a given allele will increase in the population if and only if $rb > c$.

This, then, is the general version of Hamilton' Rule. At this point, it is worth pausing and reflecting on what the variables in the general version actually mean. Because there is no such thing as a regression coefficient for a single data point, $r, b,$ and c are not actually properties of individual organisms or of any given social interaction (as a verbal description of Hamilton's Rule may lead you to believe). Instead, they are population statistics: r is the slope of the line of best fit plotting $\hat{p}_i$ against p_i for every individual in the population, and b and c can be calculated from the plane of best fit after adding w_i to the $\hat{p}_i$ against p_i plot. This formulation may at first seem strange, but it leads to a flexibility that can be either a great strength or a great weakness depending on what you want the model to achieve.

assumptions (garbage in, garbage out, to be crass), and one can argue that because the assumptions of the general version of Hamilton's Rule are so minimal, so are the scientific insights that can be gained from it.

These criticisms are all fair and not to be dismissed lightly. The general version of Hamilton's Rule is in fact of little help in predicting evolutionary change in a given ecological situation. Even so, $rb > c$ also offers a really simple framework for empirical workers to use as starting point, without the need for more sophisticated and specific models. More exact predictions require further assumptions motivated by the particular biological system modelled. Moreover, because regression coefficients are used to define the parameters, it can only ever identify correlations, not actual causations.

There is a sense, however, in which these criticisms measure the value of the general version of Hamilton's Rule against the wrong yardstick. The point of the general version is not to provide exact fine-grained causal explanations of particular scenarios, but to instead offer a way to classify, compare, and contrast more detailed causally appropriate models (the kind that biologists have been making for decades; Gardner et al. 2007, 2011; Marshall 2015). As an organizing framework, the general version also offers a way to identify common mechanisms in the origin of social behaviours. It provides a classificatory scheme and common vocabulary to translate among models, which allows more detailed theoretical models to be interpreted, compared, and contrasted in a unified way. Having mathematical frameworks that themselves make no predictions but facilitate comparisons among more specific models is in no way unique to social evolution. For example, this also is the case in physics, as recently expressed by Nobel Prize winner Steven Weinberg: 'Our most important theories, like Newtonian mechanics and quantum mechanics, are not falsifiable, because they do not make predictions by themselves, but provide general frameworks for more specific theories, which do make predictions' (quoted in Horgan 2015). By this yardstick, Hamilton's Rule can be said to be doing rather well. For example, while the mathematical details of Axelrod and Hamilton's game theoretical 'tit-for-tat'-models (Axelrod and Hamilton 1981) and Taylor and Frank's kin selection differential equations (Taylor and Frank 1996) are quite different, they can both be can be translated into the general framework and shown to satisfy the $rb > c$ inequality.

The price of this generality is that in such comparisons r, b, and c, are no longer meaningful separate entities. Rather, what matters is c and the compound entity rb. Should this be called Hamilton's Rule?

Would it not be more accurate to rename *rb* as a single, rather than a compound entity? One might say that still calling this Hamilton's Rule, despite the intuitive notions of cost and benefit no longer residing in *b* and *c*, is an artificial attempt to keep an old notation that is no longer relevant. Whatever one calls it, the general version of Hamilton's Rule does make it possible to see biological phenomena as disparate as limited dispersal, shared habitat preferences, kin recognition, greenbeards, and horizontal gene transfer all as mechanisms that can generate a positive genetic correlation between actor and recipient of a given social behaviour. By the same token, punishment, reciprocity, linkage, and pleiotropy are all mechanisms that link behaviours that are costly in the short-term but that have a long-term direct fitness benefit.

The value one assigns to grouping such different biological phenomena may differ. If one is primarily interested in specific adaptations at the organismal level, they may be so different as to make their grouping uninteresting. Under the gene's-eye view, the central question is typically a version of 'under these circumstances, will gene G be selected for or against?' By this view, the above unification is in no way a trivial achievement.

4.3 The gene's-eye view and inclusive fitness: equivalence or historical accident?

Both Williams and Dawkins saw a great deal of value in the concept of inclusive fitness and both repeatedly stressed its equivalence with the gene's-eye view. Williams even came tantalizingly close to discovering kin selection in 1957 in a paper with his wife Doris (Williams and Williams 1957). In *Adaptation and Natural Selection*, published only a few years after Hamilton's key papers, Williams wrote:

> Genic selection should be assumed to imply the current conception of natural selection often termed neo-Darwinian. An organic adaptation would be a mechanism designed to promote the success of an individual

organism, as measured by the extent to which it contributes genes to later generations of the population of which it is a member. It has the individual's inclusive fitness as its goal.   (Williams 1966, pp. 96–97).

Thus, genic selection and organisms acting to maximize their inclusive fitness are two sides of the same coin. This point has also been made by Dawkins several times. He once defined (with Hamilton's approval, as he was careful to point out) inclusive fitness as 'that property of an individual organism which will appear to be maximized when what is really being maximized is gene survival' (Dawkins 1978; see also Dawkins 2015a, p. 318; Table 4.1).

Table 4.1. The equivalence between genic selection and inclusive fitness

Unit of selection	Role	Quantity maximized
Gene	Replicator	Survival
Individual organism	Vehicle	Inclusive fitness

Hamilton himself often used a gene's-eye view early in his career. After his 1963 and 1964 publications on social behaviour, he moved on to thinking about sex ratios. In his 1967 paper 'Extraordinary sex ratios' in *Science*, he tackled Fisher's argument for why the sex ratio in most species is 1:1 (Hamilton 1967). Fisher's argument, which has been described as 'probably the most celebrated argument in evolutionary biology' (Edwards 1998), goes as follows. Suppose that one of the sexes is rarer than the other. Any parent who can adjust the sex of their offspring towards that of the rarer sex will now produce offspring with higher fitness. Genes leading to the production of the rarer sex will therefore spread, but the selective benefit of these genes will diminish the closer the population gets to 1:1.

Hamilton demonstrated that Fisher's argument does not hold equally for all genes. In the heterogametic sex, males in XY systems and females in ZW systems, the sex determining chromosome (Y or W) does not 'care' about the homogametic sex. As a consequence, if a gene that affects the sex ratio of the offspring that a parent produces is located on the Y or W it can result in deviations from the 1:1 ratio, what Hamilton called 'extraordinary' sex ratios. Hamilton's point highlights the potential for conflict among genes over sex ratio. Genomic conflict is arguably a phenomenon that makes most sense from the gene's-eye view, but it was largely a theoretical speculation at the time Hamilton was writing. By now, the study of genomic conflicts is a thriving field, and how the gene's-eye view helps make sense of such conflicts is a major theme of Chapter 5.

Hamilton was also very willing to publicly speak up for the gene's-eye view. As mentioned earlier, he wrote a very enthusiastic review of *The Selfish Gene* for *Science* (Hamilton 1977a) and in a letter to the editor forcibly protested Richard Lewontin's critical review for *Nature* of the same book (Lewontin 1977a; Hamilton 1977b). On the other hand, he seems to have preferred the individual-centred inclusive fitness concept over the gene's-eye view. Take this passage from his comprehensive review of altruism in social insects:

> A gene is being favored in natural selection if the aggregate of its replicas forms an increasing fraction of the total gene pool. We are going to be concerned with genes supposed to affect the social behavior of their bearers, so let us try to make the argument more vivid by attributing to the genes, temporarily, intelligence and a certain freedom of choice. Imagine that a gene is considering the problem of increasing the number of its replicas... (Hamilton 1972)

An argument straight out of the gene's-eye view-playbook! Yet, only a few paragraphs later he writes: 'We can now abandon the fanciful viewpoint of individual genes' and writes the rest of the paper from the perspective of the inclusive fitness interests of the individual organism. It should be emphasized that Hamilton never explicitly rejected the gene's-eye view. As far as I can tell, he considered the gene's-eye view and inclusive fitness to be fully compatible, two interpretations of the Necker cube (Table 4.1).

Nevertheless, the failure to fully commit to his own conceptual revolution appear to have frustrated some of his biggest supporters. Both Richard Dawkins and John Maynard Smith viewed inclusive fitness a sound concept, but unnecessarily messy and prone to misunderstandings. 'Paradoxically, the logical conclusion to his ideas should be the eventual abandonment of his central concept of inclusive fitness', as Dawkins (1978) put it. Both thought that Hamilton introduced the concept to save the individual as the central player in evolution and so did not follow through on his own logic and embrace a gene's-eye view. In *The Extended Phenotype*, Dawkins wrote: 'Historically, indeed, I see the concept of inclusive fitness as the instrument of a brilliant last-ditch rescue attempt, an attempt to save the individual organism as the level at which we think about natural selection' (Dawkins 1982a, p. 187). In *Brief Candle in the Dark*, Dawkins, in characteristic prose, describes inclusive fitness as a 'regrettably cumbersome bending over backwards to rescue the individual as the focus of our Darwinian attention instead of the gene' (Dawkins 2015a, p. 319). Maynard Smith, in equally characteristic prose, simply called it 'an absolute swine to calculate' and expressed his bafflement as to why Hamilton moved away from his genic explanation of the 1963 *American Naturalist* paper only to present 'a very, very, much more difficult model' (Maynard Smith 1997).

For the past decade or two, Dawkins has only rarely intervened in evolutionary debates. One issue that has proved the exception has been inclusive fitness and group selection. For example, in 2008 he wrote to the *New Scientist* to register his displeasure at an article outlining E.O. Wilson's new-found rejection of kin selection in favour of group selection (Fanelli 2008; Dawkins 2008b). Wilson had been an early supporter of Hamilton's work (though inclusive fitness is misdefined in *Sociobiology*, as Grafen 1982 has pointed out). Towards the end of his career, however, he came out strongly against it. The *New Scientist* piece largely stemmed from a paper Wilson wrote for *Bioscience*, where he used examples from eusocial insects to argue against inclusive fitness theory (Wilson 2008). Then, 2 years later, he teamed up with Tarnita and Nowak to publish the theory-heavy 'The evolution of eusociality' (Nowak et al. 2010). At a plenary talk during the 2016 meeting for The International Society of Behavioural

Ecology in Exeter, UK, Dawkins would say that *Nature*'s decision to publish Nowak et al. (2010) had turned the journal into the 'National Enquirer of science'.

In 2012, Dawkins also published two short essays going hard against Wilson's conversion. One was a review of Wilson's book length version of his argument (Wilson 2012), which Dawkins dismissed using the popular witticism 'not a book to be tossed lightly aside. It should be thrown with great force' (Dawkins 2012a). The other was a more positive commentary on Steven Pinker's 'The false allure of group selection' (Pinker 2012), where he reiterated his call for talking about replicators and vehicles instead of individual and group selection (Dawkins 2012b).

Group selection thus still appears to be the button on the back of Dawkins's head that can be most effectively pushed. Dawkins and the modern-day critics of inclusive fitness may, however, have more in common than they would probably like to admit. Take the following two quotes:

> What matters is gene selection. All we need ask of a purportedly adaptive trait is, 'What makes a gene for that trait increase in frequency? (Author 1)

> We may rely on a straightforward genetic approach: Consider mutations that modify behavior. Under which conditions are these mutations favored (or disfavored) by natural selection? The target of selection is not the individual, but the allele or the genomic ensemble that affects behavior. (Author 2)

Both are arguing against thinking in terms of individuals and/or groups, and for focusing on gene level selection. The first is from Dawkins (2008b) critiquing Wilson's group selectionism and the second from Allen et al. (2013) where Nowak and Wilson were joined by Benjamin Allen to describe the limitations of inclusive fitness.

The complaint that inclusive fitness is unnecessarily complicated, prone to misunderstanding, and that one should just focus on whether an allele with a given effect on social behaviour will invade the population is thus shared by unlikely bedfellows. When speaking at the Oxford Union, a debating society, in February 2014, Dawkins conceded that Nowak and colleagues are correct that inclusive fitness is often very difficult to calculate in practice (Dawkins 2014). At the same time, he insisted that inclusive fitness follows deductively from population genetics and that the way to prove it is not through experiment, but logically, just as one would not prove Pythagoras's Theorem by measuring angles with a ruler in nature. Dawkins concludes his reasoning by saying that he would prefer to 'forget about inclusive fitness and go straight to the level of the gene'. Nowak and colleagues are in this way joining a longstanding critique of inclusive fitness.

Part of the problem could be that there appears to be some confusion among the authors of Nowak et al. (2010) about whether or not their model is actually a critique of the gene's-eye view. The main paper is largely a group-selection flavoured verbal model, whereas the meaty appendix relies on a gene-centric mathematical approach. Following the paper's publication, the authors have also given somewhat contradictory statements. For example, in an interview with *The Guardian*, a British broadsheet, E.O. Wilson said of selfish gene thinking: 'I have abandoned it and I think most serious scientists working on it have abandoned it. Some defenders may be out there, but they have been relatively or completely silenced since our major paper [Nowak et al. 2010] came out' (Johnston 2014). At the same time, Nowak, Tarnita, and Wilson write in the 2010 paper: 'A "gene-centered" approach for the evolution of eusociality makes inclusive fitness theory unnecessary' (Nowak et al. 2010). A similar point is reiterated in a statement signed by all three authors on website of Harvard's Program for Evolutionary Dynamics, where Nowak is the director: 'our model for the evolution of eusociality is not a group selection model; instead it describes selection operating at the level of genes' (Nowak et al. 2011). Thus, critics of Hamilton's work like, at least some of the time, to frame their argument in terms of gene selectionism. This kind of rhetoric is also evident in Nowak's contribution on inclusive fitness to the edited volume *This Idea Must Die* (Nowak 2015). Demonstrating the difficulty (some would say futility) in trying to sort opinions under clear labels, Nowak himself resists the term 'gene's-eye view' preferring 'population genetics' or 'evolutionary dynamics' (M.A. Nowak, personal communication).

4.3.1 Formal connections between the gene's-eye view and Hamilton's Rule

What, then, are the actual formal connections between the gene's-eye view and Hamilton's Rule? Andy Gardner, one of the most enthusiastic defenders of Hamilton's work, has questioned the commonly held view that the two are intimately linked. In the fun 'Kin selection under blending inheritance', Gardner addresses the suggestion that the reason why Darwin failed to develop a proper theory of kin selection was that he lacked a functional theory of inheritance (Gardner 2011). He points out that because the current version of Hamilton's rule relies on the Price equation, which can handle any form of inheritance, it does not require discrete Mendelian inheritance. A breeding value does not necessarily have to be calculated from the weighted sum of allelic values.

By re-deriving the general version of Hamilton's Rule without assuming any particular inheritance, Gardner shows that blending inheritance complicates the computation of relatedness coefficients but does not in any way make their calculation impossible. Crucially, predictions remain the same regardless whether inheritance is Mendelian or blending. Based on this, Gardner argues that the individual organism is the unit of selection and adaptation, and that the gene's-eye view will only play this role in limited cases, such as those involving genomic conflicts. According to this argument, the strong connection between the gene's-eye and Hamilton's Rule stems less from formal links and more from their historical association.

There are other reasons, however, to view the two as being connected in a formal, though subtle, way. To see why, let's revisit the derivation of Hamilton's Rule outlined in Box 4.1 and explore how the gene's-eye view fits in. The derivation shows why the general version of Hamilton's Rule works best under a genic view of the environment. Focusing on $Cov(w_i, p_i)$ allows the natural selection part of evolutionary change to be captured, but only in a constant environment. Making such an assumption requires us to think, as Fisher did, of the environment of a given allele to include not just the external environment, but also the rest of the genome (Fisher 1918). From the perspective of an individual organism, conceptualizing any changes in the genomic context (other genes in the genome and in the gene pool) that an allele may find itself in as environmental makes little sense. From the gene's-eye view it is straightforward. The only way that $Cov(w_i, p_i)$ can be taken to capture natural selection in a constant environment is if this gene's-eye view of the environment is adopted. The connection between inclusive fitness theory and the gene's-eye view may appear to be a subtle point in the derivation of the general version of Hamilton's Rule, but it demonstrates that the intimate relationship is not just a historical accident.

Examining the general version of Hamilton's Rule also opens an interesting door to some potential tensions with the gene's-eye view. Recall that a key step in Queller's derivation of the general version of Hamilton's Rule is that the transmission term, $E(w_i \Delta p_i)$, of the Price equation is assumed to be 0. This means that genetic drift, among other things, is ignored, which is usually acceptable if the primary concern is adaptation. From a gene's-eye view, a more important consequence is that it forces us to ignore within-organism selection due to, for example, meiotic drive or other forms of genomic conflict. This is a rather high price to pay. That all genes work together for the same purpose is a crucial assumption of many inclusive fitness models. The gene's-eye view, however, stresses that such unity of purpose cannot just be taken for granted. It is there and it must be explained.

To see whether this potential tension actually manifests, I next examine the tradition in evolutionary biology that has sought to use inclusive fitness as the answer to the question of what is it that organisms should appear designed to be trying to maximize. I therefore turn to this tradition's most ambitious version, Alan Grafen's Formal Darwinism Project.

4.4 Maximization of inclusive fitness and the Formal Darwinism Project

The Formal Darwinism Project takes as its starting point what it considers to be Darwin's central insight: that evolution by natural selection provides a way by which a purely mechanistic process (inheritance and reproduction) can give rise to the appearance of design in the living world (see also Dennett 1995 and Haig 2020). Its goal is to formalize this insight. The chief architect of this unification effort is Alan Grafen, who received his undergraduate degree in experimental psychology and a master's degree in economics before doing his doctorate in evolutionary biology under Dawkins's supervision. Together with Andy Gardner, he is probably the best representative of the contemporary neo-Paleyan tradition in British biology introduced in Chapter 1.

Grafen seeks to construct a mathematical bridge between the dynamic models of population genetics that capture changes in gene frequencies and the optimality models that describe fitness maximization. By doing so, he wants to mathematically justify the 'individual-as-maximizing-agent analogy', which he argues is an analogy taken for granted by many empirical biologists, especially behavioural ecologists, but is one that lacks formal grounding. Grafen's bridge-building attempt began with Grafen (1999) and now spans several papers, many abstract and technical even for biologists with a serious theoretical bent. Grafen (2007) and Grafen (2008) offer non-mathematical overviews and the 2014 special issue of *Biology and Philosophy* provides an introduction to the many conceptual issues of the project (see the opening paper by the editors Okasha and Paternotte 2014 and Grafen's outline and response to the eleven commenting papers; Grafen 2014a; 2014b).

The Formal Darwinism Project is the most comprehensive expression of a broader tradition that views inclusive fitness as the answer to the design objective of natural selection, i.e. what organisms should appear to maximize. If both defenders and critics agree that there are some mathematical shortcomings to inclusive fitness, they disagree over how comfortable one can be with them. Key to predicting whether someone can live with these limitations seems to be whether they share the neo-Paleyan assumption underlying the Formal Darwinism Project. Authors like Allen and Nowak firmly reject this notion and see no need for any kind of design principle (Allen et al. 2013; Allen and Nowak 2016).

Those who do take adaptation to be a special problem are more likely to be comfortable with the assumptions required to make models of maximizing inclusive fitness work (West and Gardner 2013; Marshall 2015; Levin and Grafen 2019). For example, such models often assume that selection is weak and that the fitness effects of social behaviour are additive. Advocates of inclusive fitness concede these points and instead emphasize another aspect of the general version of Hamilton's Rule. Just as the point of Hamilton's Rule is to provide an organizing framework rather than exact testable predictions for every conceivable biological scenario, inclusive fitness is not just another mathematical approach. Instead, it explains what organisms should appear designed to act as if to maximize (West and Gardner 2013; Levin and Grafen 2019; see also Lehmann and Rousset 2020; Paternotte 2020). From this, it also follows that cooperation and conflict can be modelled as situations where the inclusive fitness interests align or diverge, which is why it can be such a powerful and popular tool in social evolution.

4.4.1 Genes versus individuals in the Formal Darwinism Project

There has always been a connection between the gene's-eye view and the Formal Darwinism Project. In 2011, when Grafen teamed up with the mathematician Charles Batty, like Grafen a fellow at St John's College, Oxford, to advertise for two postdoc positions in mathematics to work on the Formal Darwinism Project, the project was advertised with the title 'The deep mathematical theory of selfish genes'. As with the gene's-eye view's general relationship with inclusive fitness, however, there is also some tension. Recall that the existence of such tension is perhaps not surprising. Dawkins warned of the 'theoretical dangers' associated with thinking in terms of individuals maximizing their fitness (Dawkins 1982a, p. 91) and he further wrote of Hamilton that:

> instead of following his ideas through to their logical conclusion and sweeping the individual organism from its pedestal as notional agent of maximization, he exerted his genius in devising a means of rescuing the individual. He could have persisted in saying: gene survival is what matters; let us examine what a gene would have to do in order to propagate copies of itself. Instead he, in effect, said: gene survival is what matters; what is the minimum change we have to make to our old view of what individuals must do, in order that we may cling on to our idea of the individual as the unit of action?   (Dawkins 1982a, p. 196)

Grafen long emphasized that 'the organismal approach is not in conflict with the "gene selectionism" of Dawkins' (Grafen 1984). In the early papers of the Formal Darwinism Project, Grafen explicitly framed the aims of the Formal Darwinism Project in terms of formalizing insights from Dawkins and Williams (Grafen 1999, 2002, 2007). In more recent papers, the gene's-eye view has taken a back seat. For example, Levin and Grafen (2019) write that 'the selfish gene approach can be useful for certain gene level questions, such as intragenomic conflict'. If one grants the premise of the Formal Darwinism Project— that Darwin unified mechanism and purpose and that the appearance of design is at the heart of that—why should the effort to formalize this unification privilege an individual over a gene-centric perspective?

One way to think about it is that only organisms can be said to manifest complex adaptations. Gardner (2014b) makes this point by appealing to Paley's criteria for complex design: contrivance and relation (see Chapter 1). Genes, he argues, satisfy one of the two criteria, organisms satisfy both. Contrivance captures how an entity appears designed for a specific purpose. Genes can satisfy this requirement, whether it is to produce a certain protein or to regulate the expression of other genes. Even if genes can have many functions, however, they do not come near the repertoire of traits and functions that a whole organism can manifest. They therefore fail to meet the 'relation of parts' requirement.

How parts of the organisms work together is also key to another potential tension point between genic and organismal approaches to adaptation. How should conflict among genes be handled? Hamilton captured the issue rather poetically:

> Seemingly inescapable conflict within diploid organisms came to me as both a new agonizing challenge and at the same time as a release from a personal problem I had had all my life (...) Given my realization of an eternal disquiet within, couldn't I feel better about my own inability to be consistent in what I was doing, about indecision in matters ranging from daily trivialities up to the very nature of right and wrong? (...) As I write these words, even as to be able to write them, I am pretending to a unity that, deep inside myself, I now know does not exist. I am fundamentally mixed, male with female, parent with offspring, warring segments of chromosomes...   (Hamilton 1996, pp. 134–134).

There is an argument that for an entity to be able to evolve adaptation, all parts of that entity must work towards the same goal and there can be little to no selection within that entity (Gardner and Grafen 2009; West and Gardner 2013; Gardner 2014a, 2014b; Grafen 2014a). That is, the parts must demonstrate a unity of purpose. At the gene level, this means no genomic conflict. Recall that the derivation of the general form of Hamilton's Rule comes with an assumption of perfect genetic transmission, i.e. fair meiosis such that $E(w_i\Delta p_i)$ equals 0, and so no role for segregation distorters and other kinds of selfish genetic elements (Box 4.1). Grafen's Formal Darwinism models make the same assumption.

Selfish genetic elements have the potential to undermine the unity of purpose required for this assumption. Their existence raises the question whether it is appropriate to treat the individual organism as the sole fitness maximizer (Okasha 2018, p. 29). As Dawkins put it: 'There is a sense in which a "vehicle" is worthy of the name in inverse proportion to the number of outlaw replicators that it contains' (Dawkins 1982a, p. 134). Here, it is worth emphasizing that it is not that the Formal Darwinism Project has nothing to say about genomic conflicts. Whereas Grafen (2014b) admits that he has often tended to downplay their significance, genomic conflicts can be studied using its methods. Because it links replicator dynamics to maximization principles, its models can be applied to different part of the genome (say maternally inherited mitochondrial genes versus biparentally inherited autosomal genes), which allows an account of genomic conflicts to be developed (see Gardner and Welch 2011 and Gardner and Úbeda 2017). Rather, the implicit assumption is that if the goal is to understand organismal adaptations, such conflicts can be ignored. Whether such an assumption is reasonable is ultimately an empirical question.

Inclusive fitness theorists have generally relied on the so-called 'phenotypic gambit'. This is the assumption that selection on a trait will not be constrained by its genetic architecture (Grafen 1984). While recognizing that this assumption will often be violated, such as in the case of heterozygote advantage, it is suggested that it is justified because of its track record making empirically testable predictions (Grafen 2014a; Levin and Grafen 2019). Theoretically, the assumption is handled by assuming that the trait in question is captured by a haploid population genetic model with a separate allele for each possible phenotype. When such genic consensus cannot be achieved, the phenotypic gambit fails. This may happen, for example, when a single trait is affected by genes inherited in different ways, say by both maternally inherited mitochondrial genes and biparentally inherited autosomal genes. Such transmission asymmetries are a common source of genomic conflict (Cosmides and Tooby 1981; Burt and

Trivers 2006; Ågren and Clark 2018) and may result in the fragmenta-
tion of individual fitness maximization.

Haig has argued that the ubiquity of genomic conflict leaves biologists interested in adaptation with two possible options (Haig 2014).
The first is the one offered by the Formal Darwinism Project, which
suggests that such conflicts can typically be ignored. From a gene's-
eye view this goes too far. Under the second solution, the one
advocated by Haig, the individual organism is viewed as an
'adaptive compromise', rather than a unified fitness maximizer
(Haig 2006b, 2014). As Maynard Smith argued, the gene's-eye view
forces us to think about what individual organisms are to begin with
and why and how it achieves its unity of purpose (Maynard
Smith 1982a, 1985). As he put it: 'How did it come about that most
genes, most of the time, play fair, so that a gene's success depends only
on the success of the individual that carries it?' (Maynard Smith 1985)

Maynard Smith would return to this question when he published
The Major Transitions in Evolution together with Eörs Szathmáry
(Maynard Smith and Szathmáry 1995). As outlined in Chapter 2, they
put forward the idea that life's hierarchy is a consequence of a series
of events where new entities evolved by the coming-together of
entities that were previously surviving and reproducing on their own.
A crucial step in these transitions is the suppression of conflict at
lower levels. This process is never complete, and the unity of the individual is therefore constantly threatened from within. Genomic conflicts are not just a curiosity with little evolutionary significance but
exactly what should be expected.

The gene's-eye view and the Formal Darwinism Project share the
neo-Paleyan commitment to adaptation being the main question that
evolutionary theories should aim to explain. One difference is which
of Lloyd's four questions that they are primarily interested in. Whereas
the Formal Darwinism Project is concerned with a combination of
the interactor and manifestor of adaptation questions, the gene's-eye
view is focused on the beneficiary question. The major transitions
framework can provide a productive bridge between the individual-

as-maximizing-agent analogy and the gene's-eye view, as advocated by Bourke (Bourke 2011, 2014). The gene's-eye view, especially when paired with a major transitions framework, drums home that this analogy is conditional on the unity of purpose of genes.

4.5 Recent reconciliations between the gene's-eye view and inclusive fitness

This brief historical sketch demonstrates the intimate relationship between the gene's-eye view and inclusive fitness. The ties between the two perspectives are deep and when frustrations have arisen it has usually been from the side of the gene's-eye view, as exemplified by the comments from Maynard Smith and Dawkins. My impression is that most biologists are happy to keep both perspectives around, using either depending on the goal of the moment. As Krebs and Davies put it in one of their classic textbooks:

> 'the field biologist sees *individuals* dying, surviving and reproducing; but the evolutionary consequence is that the frequencies of *genes* change. Therefore the field biologists tend to think in terms of individual selection whilst the theorists thinks in terms of selfish genes.'   (Krebs and Davies 1993, p. 375; original emphasis)

This pluralistic pragmatic approach is very sensible. Different problems will often demand different tools. That being said, exploring how the two perspectives are related is interesting and worthwhile in of itself. There are numerous papers on inclusive fitness published every year, by theorists and empiricists, and by critics and supporters. Here, I am less interested in what this new work can tell us about inclusive fitness theory than what light it can shine on the gene's-eye view. Two recent papers on inclusive fitness theory explicitly frame their argument in terms of the gene's-eye view (Akçay and Van Cleve 2016 and Fromhage and Jennions 2019) and I will end this chapter by discussing them.

Both papers touch an especially contentious point in the current debate over inclusive fitness, which is whether inclusive fitness is a causal property that belongs to the individual organism. Hamilton was clear that this was his definition, and supporters of inclusive fitness have made much of the fact that inclusive fitness is a unique definition of fitness in that it is the only one that is both a target of selection and under the organism's control. Causal control is important because they argue that an entity can only appear designed for something that it has full control over, which makes inclusive fitness special even relative to closely related concepts, such as neighbour modulated fitness (Gardner et al. 2011; West and Gardner 2013; Marshall 2015; Levin and Grafen 2019). Again, critics deny this point (Allen and Nowak 2016).

Akçay and Van Cleve (2016) and Fromhage and Jennions (2019) both touch on the causal property issue but suggest rather different solutions and, in their own ways, challenge Hamilton's original definition of inclusive fitness. Akçay and Van Cleve reject Hamilton's argument that inclusive fitness is an extension of classical fitness and suggest that it does not actually belong to the individual but that it is a property of the genetic lineage, that is all copies stemming from an original mutation. Fromhage and Jennions, in turn, suggest re-defining inclusive fitness to align with what they call the folk definition of inclusive fitness. Such a definition, they argue, would allow inclusive fitness to be properly reconciled with a gene's-eye view.

4.5.1 The genetic lineage view of inclusive fitness

The central argument of Akçay and Van Cleve (2016) is that many of the potential pitfalls of inclusive fitness theory can be avoided if inclusive fitness is re-conceived as a property of the genetic lineage rather than of the individual organism. They base their conclusion on a pretty straightforward population genetic argument, where they consider a haploid population and ask when a rare mutation can invade the population. This kind of fitness, called invasion fitness, is in its most basic form the geometric mean growth rate of a rare mutation in a population. While this quantity, which can be denoted g, captures all that is needed to predict the long-term fate of the mutation over evolutionary time (it will invade as long as $g > 1$), it does not necessarily lend itself to much of a biological interpretation.

For a fitness definition like this to have meaning for empirically minded biologists, whether in the lab or in the field, it needs to be translated into more concrete terms. In practice, this has often meant number of surviving offspring per individual per lifetime. Denoted w, personal fitness is typically calculated by theorists as the average number of surviving offspring of a given genotype. Because any potential stochastic effects stemming from biotic or abiotic factors can be accounted for by averaging in a certain way, the remaining variation in w will come from the focal individual's social partners. By explicitly considering these interactions, Akçay and Van Cleve point out that individual and inclusive fitness can be linked so that the fitness of an individual carrying the rare mutation, w_m, can be expressed as a linear function of how common the rare mutation is among the social partners of the focal individual carrying the rare mutation, p_n:

$$w_m = 1 - c + bp_n + e$$

Where 1 is the baseline fitness of the non-mutant genotype; c is the direct fitness effect, the effect of the mutation on the fitness of the individual that carries it, and b is the indirect fitness effect, how other individuals affect the fitness of the individual that carries the mutation. e is an error term. The next step is to determine the expected fitness of the mutant across all social partners, which is achieved by considering all possible frequencies of the rare mutation among social partners. This expected term $E[w_m]$ will be the inclusive fitness, w_{IF} such that:

$$E[w_m] = w_{IF} = 1 - c + br.$$

where r here is defined as the expectation that a random social partner will be of the same mutant lineage. If fitness effects are additive, this definition is the same as Hamilton's original definition of inclusive fitness. In this model, the value of b and c do not depend on the frequency of the rare mutation, and the mutant's fitness can be captured by a linear model. This fact allows w_{IF} to be linked to invasion fitness so that the rate at which the mutation is spreading is the inclusive fitness of that mutation lineage. In other words:

$$g = w_{IF} = 1 - c + br$$

Since a rare mutation will invade when $g > 1$, the above expression can be rearranged to that say that the rare mutation will spread when the inequality $rb > c$ is satisfied. Akçay and Van Cleve thus recover Hamilton's Rule while connecting the simple theoretical concept of invasion fitness, g, to a more practical concept of individual fitness, w.

Akçay and Van Cleve's definition of the inclusive fitness of a gene comes with two key implications. The first is that it no longer makes sense to think of inclusive fitness as a generalization of classical fitness, as Hamilton originally did. Inclusive fitness is simply the fitness of a genetic lineage averaged across all possible states the lineage may find itself in, including genetic backgrounds, population structures, and biotic and abiotic environments. The second implication is that it transfers inclusive fitness from the individual organism to the genetic lineage. As discussed, the fact that inclusive fitness is under the full control of the individual organism is often cited as a key advantage of inclusive fitness theory. Relinquishing this may therefore be a big pill to swallow. Whereas Dawkins and Williams never used these exact terms, this move from organism to genetic lineage should be easier to accept for proponents of the gene's-eye view.

4.5.2 The folk definition of inclusive fitness and the parliament of genes

The goal of Fromhage and Jennions (2019) was to 'tidy up' the gene's-eye view by resolving its tension with inclusive fitness models centred on individuals. Such tidying is necessary, they argue, because there is a danger in focusing exclusively on genes. Despite the impressive precision with which theoretical population geneticists can construct mathematical models describing the conditions under which an allele with certain hypothetical properties is favoured by natural selection, the fact remains that many evolutionary biologists are usually more interested in the question of whether a given phenotype is adaptive or not. As helpful as the gene's-eye view can be for working out the logic of natural selection, it will always be impotent unless it can be properly connected to traits of actual organisms (Hammerstein 1996).

To get around this, Fromhage and Jennions suggest that the current definition of inclusive fitness should be discarded. In its place, they want the field to adopt a more intuitive version that they dub 'folk inclusive fitness', a version long repudiated by inclusive fitness theorists (e.g. Grafen 1982). This is quite a radical step (see Queller 2019 for a primer), especially as they argue that their definition can handle two of the main criticisms levelled at inclusive fitness: non-additive fitness interactions and mutations of various effect sizes.

Their new definition of inclusive fitness is a product of changing the way it is calculated. Recall from earlier that to properly calculate an individual's inclusive fitness it has been considered crucial to deduct the part of the direct fitness that is due to actions of the social environment. Fromhage and Jennions suggest that this subtraction step can be dropped, leaving what they call a 'folk definition' of inclusive fitness. The flipside of this definition, they argue, is that it makes inclusive fitness a quantity that when maximized also results in what is best for the 'majority interest' of the genome, thus linking a genic with an individual centred perspective.

Fromhage and Jennions' account seeks to combine Dawkins's vehicle concept with Egbert Leigh's 'parliament of genes' (Leigh 1971). Leigh introduced the idea of a parliament of genes in the context of the evolution of fair meiosis, i.e. that each allele in a diploid organism has a 50% chance of being transmitted. Discussing the potential selective advantage of a meiotic driver, a gene promoting its own transmission beyond 50%, he wrote: 'It is as if we had to do with a parliament of genes: each acts in its own self-interest, but if its acts hurt others, they will combine together to suppress it' (Leigh 1971, p. 249). That is, even if meiotic drivers, or any other genes that can promote their own transmission at the expense of other genes, are initially favoured, their spread will be counteracted by selection at other unlinked loci, and because there will be more such loci, they will serve the 'majority interest' (see also Scott and West 2019). According to Fromhage and Jennions, the 'majority interest' of the genome is best served by producing organisms with high 'vehicle quality', as measured by the ability to produce copies of the organism's genes. More specifically, high vehicle quality is measured by counting copies of a hypothetical gene with idealized properties, a 'reference gene', which represents the genome's majority interest.

The reference gene (as is often the case with the gene's-eye view, the 'gene' in question is actually an allele) has four key properties:

1. It is carried by the focal organism.
2. It is present at a low frequency in the population.
3. Its transmission is Mendelian.
4. It is never or rarely expressed.

These four properties serve two main functions: allowing the number of reference genes to be counted (properties 1 and 2) and to guarantee that a reference gene is a good stand-in for the genome's majority interest (properties 3 and 4).

Estimating vehicle quality by the number of reference gene copies that a focal individual is causally responsible for means that it can be quantified as follows. Under sexual reproduction, the probability of a reference gene being transmitted from parent to offspring is given by s. Assuming outbreeding, $s = 0.5$. Pedigree relatedness is given by r, and the number of propagated reference gene copies is captured by the expression $s \times \Sigma(r \times \Delta_r)$, where Δ_r is the additional offspring produced by relatives with relatedness r, as caused by the focal organism.

This way of doing the sums counts all the offspring of the focal individual, and thus it does not include the 'stripping' part of

Hamilton's original definition. This 'folk definition' of inclusive fitness is therefore maximized when the expected value of $\Sigma(r \times \Delta_r)$ is maximized. Because this happens when the majority interest of the genome is satisfied, Fromhage and Jennions argue that this offers a way to handle genomic conflicts.

Fromhage and Jennions's solution to the tension between the gene's-eye view and inclusive fitness is quite different to the one offered by Akçay and Van Cleve. Fromhage and Jennions retain the organism as a central unit, but they do so while simultaneously trying to take conflicts between genes seriously in a way that the Formal Darwinism Project usually does not. Akçay and Van Cleve moved inclusive fitness from the organism down to the gene, which can be more readily accepted by those with a gene's-eye view and arguably comes close to Williams's and Dawkins's thinking. At the same time, both Williams and Dawkins did think the design-like features of organisms were something special and if our explanations completely leave them out in favour of focusing on one gene at a time something is missing.

4.6 Summary

- W.D. Hamilton provided one of the earliest clear articulations of the gene's-eye view and there is a long-standing, intimate, relationship between the gene's-eye view and the concepts of inclusive fitness, kin selection, and Hamilton's Rule.
- For pragmatic purposes, the gene's-eye view and inclusive fitness can often be thought of as equivalent, two sides of a Necker cube: genes are replicators that try to maximize their survival and organisms are vehicles maximizing their inclusive fitness. At the same time, Dawkins and Maynard Smith, both admirers of Hamilton, complained that he did not follow through on his own conceptual revolution in fully shifting from an organismal to a genic perspective.

- Inclusive fitness and Hamilton's Rule have recently come under much scrutiny. Strikingly, and somewhat ironically, today's critics of Hamilton's work often rely on the gene's-eye view's rhetoric and use arguments similar to those first articulated by Dawkins and Maynard Smith.

- The Formal Darwinism Project seeks to formally justify that individual organisms should strive to maximize their inclusive fitness by unifying population genetic and optimization models. Mathematically rather abstract, it is based on an assumption of genomic unity of purpose that highlights some interesting potential tension with the gene's-eye view.

- Despite traditionally thought of as equivalent, there is interesting contemporary work on the relationship between the gene's-eye view and inclusive fitness.

5

Empirical Implications

5.1 Introduction

When the Manhattan Project physicist Leo Szilard switched to biology after the end of the war, he found that this career change ruined his bath-time. As a physicist, he was accustomed to starting his day in the bath, while thinking through theoretical problems in his head (Barry 2005, p. 24). In his new life as a biologist, this did not work; he had to repeatedly get up to look up a fact, ruining any attempts of Archimedes-style ruminations.

Biology by its nature is messy, a science of exceptions. In many corners of biology, the dominant view of science is one where facts come first and then, in light of those, new theories emerge. In evolutionary biology at least, that is far from always the case and theory often leads the way. The early success of the gene's-eye view stemmed from helping make sense of the field's old problems, especially those related to social behaviour. It also had more radical implications.

Nowhere are these spelled out more clearly than in *The Extended Phenotype* (Dawkins 1982a). Every writer has one piece that they consider their best. Something about which they say: 'if you read only one thing of mine, read this!'. For Dawkins, that is *The Extended Phenotype*. It was his second book and it is the only book of his that is aimed at professional biologists, rather than the general public. It is centred upon the goal of '[freeing] the selfish gene from the individual

The Gene's-Eye View of Evolution. J. Arvid Ågren, Oxford University Press. © J. Arvid Ågren 2021.
DOI: 10.1093/oso/9780198862260.003.0006

organism which has been its conceptual prison' (Dawkins 1982a, p. vi). *The Extended Phenotype* remains the most exhaustive articulation and defence of the gene's-eye view. The radical nature of its argument is well illustrated by the afterword by Daniel Dennett that was added to the 1999 paperback edition:

> Is *The Extended Phenotype* science or philosophy? It is both; it is science, certainly, but it is also what philosophy should be, and only intermittently is: a scrupulously reasoned argument that opens our eyes to a new perspective, clarifying what had been murky and ill-understood, and giving us a new way of thinking about topics we thought we already understood (...) [Dawkins] shows how our traditional way of thinking about organisms should be replaced by a richer version in which the boundary between organism and environment first dissolves and then gets partially rebuilt on a deeper foundation.   (Dennett 1999)

This chapter is dedicated to three empirical phenomena whose study grew out of the gene's-eye view and constitute the core of this new deeper foundation: extended phenotypes, greenbeard genes, and selfish genetic elements.

Extended phenotypes are the effects of a gene that occur beyond the physical body in which the gene resides. Paradigmatic examples include beaver dams, brood parasitism by cuckoo chicks, and parasite manipulation of host behaviours. Greenbeard genes are genes that can recognize copies of themselves in other individuals and then cause their carrier to act nepotistically toward such individuals. Greenbeards make clear that, from a gene's-eye view, for a costly social behaviour to evolve it is the relatedness at the particular locus that underlies the social behaviour that matters, not the genome-wide relatedness. Greenbeards were initially thought to be a fun theoretical idea unlikely to exist in nature, but several examples have been identified, including in yeast, slime molds, and ants.

Selfish genetic elements are stretches of DNA that can promote their own transmission at the expense of other genes in the genome while having either no effect or a negative effect on organismal fitness. The first description of selfish genetic elements predates the origin of the gene's-eye view, but those examples were usually considered to be genetic oddities with few implications for evolutionary theory. Although selfish genetic elements played only a minor role in the early articulations of the gene's-eye view, the gene-centred perspective brought them to the forefront of evolutionary biology. Reciprocally, selfish genetic elements have often been considered the best evidence for the power of the gene's-eye view.

This chapter will be slightly different than the others. Rather than navigating the intricacies of conceptual debates, I will review empirical examples that all bolster the gene's-eye view's effort to undermine the centrality of the individual organism in evolutionary explanations. As Dawkins noted soon after the publication of *The Selfish Gene*: 'All these ideas, even if they appear far-fetched in practice, are perfectly respectable in theory, and you would never think of them if you based your ideas on individual fitness rather than on gene replication' (Dawkins 1978). Selfish genetic elements are also my own area of research and I will pay particular attention to their biology. I will end the chapter by discussing how the historical association between the gene's-eye view and these empirical examples holds up today, especially when compared to other proposed frameworks.

5.2 Extended phenotypes

Dawkins introduced the term 'extended phenotype' during a talk at the 15th International Ethological Conference in Bielefeld, Germany (later published as Dawkins 1978). In *The Extended Phenotype*, he summarized the central claim of the concept as: 'An animal's behaviour tends to maximize the survival of the genes 'for' that behaviour, whether or not those genes happen to be in the body of the particular animal performing it' (Dawkins 1982a, p. 233). Together with group selection, the concept of extended phenotypes has been a rare topic that has led Dawkins to publicly comment on current debates in evolutionary biology. Whereas the former has been a reliable trigger of irritated interventions, work on extended phenotypes has elicited more enthusiastic ones. In the foreword of a 2012 edited volume on the topic (Hughes et al. 2012), Dawkins names parasites manipulating the behaviour of their hosts as his 'personal epitome of Darwinian adaptation, the *ne plus ultra* of natural selection in all its merciless glory' (Dawkins 2012c). And in an essay for a special issue of *Biology and Philosophy* commemorating the 20th anniversary of *The Extended Phenotype*, Dawkins dreams of opening an 'Extended Phenotypic Institute'. In the dream, the institute is made up of three departments: the Zoological Artefact Museum, the Laboratory of Parasite Extended Genetics, and the Centre for Action at a Distance (Dawkins 2004b). In his autobiography he nominates the entomologist and evolutionary biologist David Hughes to be its inaugural director (Dawkins 2015a, p. 337).

At the same time, Dawkins also admits of being a little disappointed that empirical research on extended phenotypes has not made more progress. He notes that whereas the gene's-eye view is now common knowledge among professional evolutionary biologists, and some empirical fields spurred by the concept, such as the study of genomic conflicts have seen rapid developments, extended phenotypes—the part of the general theoretical framework of the gene's-eye view that he considers wholly his own—is still a minor topic. That might well be true, but several examples have been described. They can be organized into three main categories.

5.2.1 Three kinds of extended phenotypes

The wide variety of potential extended phenotypes mean that there are several ways in which they can be grouped. Here, I have chosen to follow the approach reflected in Dawkins's imaginary Extended Phenotypic Institute (see also Hughes 2008, 2012, 2013): animal architecture, action at a distance, and parasite manipulation of host behaviour.

Animal architecture has long fascinated biologists. In 1974, it was the name of a book by Karl von Frisch, who in the previous year had shared the Nobel Prize in Physiology or Medicine with Konrad Lorenz and Dawkins's doctoral advisor Nikolaas Tinbergen for their foundational work on animal behaviour (von Frisch 1974). The 1999 paperback edition of *The Extended Phenotype* came with a beaver dam on the cover and that has often been the textbook example of the concept.

The beaver dam also serves as a useful starting point for how to restrict the definition of an extended phenotype. Dawkins (2004b) describes how lay people often ask whether buildings count as extended phenotypes. The answer is no, for the fates of the buildings do not typically affect the fitness of the architects responsible. Extended phenotypes only count if their variation is correlated with the success of alleles responsible for that variation. This causal link exists for beaver dams, but not for skyscrapers. Such a causal link is also what sets extended phenotypes apart from niche construction. Niche construction as a term was coined by John Odling-Smee (Odling-Smee 1988; Odling-Smee et al. 2003), but it is perhaps best captured by Levins and Lewontin in *The Dialectical Biologist* who wrote: 'the organism influences its own evolution, by being both the object of natural selection and the creator of the conditions of that selection' (Levins and Lewontin 1985, p. 106). Niche construction thus encompasses all sorts of alterations of the environment by organisms. That is, regardless of whether they are mere by-products of an organism's life cycle with no effect on its fitness or engineered products with such effects. To proponents of niche construction (such as Laland et al. 2016), extended phenotypes are a special case of the former, whereas Dawkins considers the definition of niche construction to be so broad as to be useless (Dawkins 2004b; 2015, p. 336). In part, this difference in opinion reflects the gene's-eye view's stronger focus on explaining adaptations. In part, the attempt to elevate niche construction theory is associated with a broader criticism of much standard evolutionary theory. I am not going to adjudicate that difference here. Papers in the *Biology and Philosophy* special issue mentioned above (Dawkins 2004b; Laland 2004; Turner 2004; Jablonka 2004) and the more recent exchange in *The Journal of Genetics* (Gupta et al. 2017a, 2017b; Feldman et al. 2017) provide some flavour of the kind of emotions that that debate elicits.

Even within the stricter definition of extended phenotypes, there are plenty of examples of animal architecture. Mike Hansell documents several of them and show that they typically fulfil one of three functions: providing a secure home to live in, capturing prey, and communicating with members of the same species (Hansell 2005). Hansell's comprehensive review demonstrates just how much is known about how and why animals build. Yet, Dawkins (2004b) lamented that twenty years after *The Extended Phenotype,* nobody had studied the genetics of animal architecture.

A decade later, such a study finally appeared. Weber et al. (2013) showed that the shape and size of the escape tunnels used by oldfield mice (*Peromyscus polionotus*) are associated with a surprisingly small number of genes. Using a quantitative trait locus mapping approach, they identified four independent regions of the genome, each on a different chromosome, that are associated with variation in the tunnel phenotype. One genomic region correlates with whether an individual builds a tunnel at all, and three control the size of the tunnel, each responsible for about 3 centimetres of length. The study beautifully illustrates how the gene's-eye view re-imagines the boundary between organism and environment. Just as the environment is not restricted to the outside of the organism, the phenotypic effects of a gene are not limited to the organism itself.

Dawkins's second kind of extended phenotype, action at a distance, refers the manipulation of behaviour by another individual, of the same or different species. To Dawkins, the theoretical underpinnings of action at distance can be traced to a paper he co-authored with John Krebs on animal signals (Dawkins and Krebs 1978; see also Krebs and Dawkins 1984). In it, they reject the field's previous interpretation of animal signals as always being cooperative, where cooperation is aided by shared information. As an alternative hypothesis, they argued that many signals are best thought of as attempts to manipulate the behaviour of the receiver. The idea of manipulation *per se* was not new (already Aristotle knew about the brood parasitism of the European Cuckoo; Schulze-Hagen et al. 2008), but the paper was part of the general shift towards an increased focus on conflict in the study of social behaviour.

Action at a distance may take many shapes and forms. One is that of brood parasitism in birds, which occurs when a bird lays its eggs in the nest of another individual (Davies 2000). Brood parasites reap the rewards of parental care, including incubation and care of nestlings, without having to pay the associated costs. Brood parasitism, which occurs in about 1% of bird species, often takes ingenious forms. A parasitic mother keeps a close eye on the host's nest, making sure that her timing is exactly right to sneak her eggs in. The parasite eggs often closely resemble those of the host, and after hatching, the chicks exploit their host parents in several ways. For example, hatchlings of the European Cuckoo eject host eggs from the nest, leaving the host with nothing but cuckoo eggs. Because the host, often a Garden Warbler or Reed Warbler, is much smaller, this results in the striking situation in which the young cuckoo quickly grows larger than its foster parent. Hosts are not completely helpless; they are often involved in a co-evolutionary arms race with the parasites, resulting in impressive changes to the colour and patterning of eggs (see Jamie 2017 for an excellent overview). Current research continues to reveal surprising results about avian brood parasitism. Its genetic architecture, however, remains poorly understood.

The situation is different in the third category of extended phenotypes: parasites manipulating the behaviour of the host in which they reside. This category shares many features with action at a distance but differs in that the manipulator is located inside of the individual being manipulated. The experimental tractability of microorganisms have meant that researchers have been able to identify a parasite gene involved in causing the change in behaviour of a host. Healthy caterpillars of the European gypsy moth spend most of their days in the soil to avoid predation, only venturing into trees to feed on leaves at night. When they are infected by the baculovirus (*Lymantria dispar* nucleo-polyhedrovirus), however, they climb to the treetop, where they die, liquefy, and release the viral particles. This behavioural shift, Hoover et al. (2011) elegantly demonstrated, is caused by the *egt* gene of the virus.

As dramatic as this viral manipulation of gypsy moth behaviour may seem, it is far from unique. Similar stories have been reported in a wide variety of systems (Moore 2002; Hughes et al. 2012). One striking example is that of rats infected by *Toxoplasma gondii*. These rats lose their usual fear of cats, even developing an attraction to the urine of their wonted predator. Once eaten, the parasite moves from the rat to the cat, where it settles down and mates (Berdoy et al. 2000). Another is that of zombie ants (Hughes et al. 2011; Fredericksen et al. 2017). The fungal parasite *Ophiocordyceps unilateralis* infects the carpenter ant host *Camponotus castaneus* and initially behaves similarly to the baculovirus in the gypsy moth. It starts by making the ant move away from the safety of its nest to a part of the plant that exposes the ant but that has the right conditions for fungal growth. The process culminates with the fungus growing a long stalk straight through the head of the ant. Because the stalk is topped with a bulb full of spores, this allows the fungus to spread further.

Adopting a gene's-eye view helps make sense of these bizarre behaviours. A gene will be selected for even if the selected effects are outside of the body in which the gene is located. They also illustrate how proponents of the gene's-eye view want to rebuild the concept of an organism: not as a physical body, but as 'an entity, all whose genes share the same stochastic expectation of the future' (Dawkins 1990). Host and parasite genes are in a way part of the same body, as they reside in the same physical structure, but they have different expectations of the future. Genes of vertically transmitted commensal microbes have expectations better aligned with those of their host organisms. Even more so, that genes of the same genome share the same expectation is often taken for granted, but two examples, greenbeard genes and selfish genetic elements, show why that is not necessarily the case.

5.3 Greenbeards

The concept of greenbeards is one of the most successful memes of the gene's-eye view and a salient example of how new theory identifies new phenomena (Grafen 1998). The name itself comes from a thought-experiment first presented by Hamilton (1964b) and then developed and given its current name by Dawkins in *The Selfish Gene* and in *The Extended Phenotype* (Dawkins 1976, pp. 115–116, 1982a, pp. 143–155). To Hamilton, the thought experiment was a way to demonstrate that his concept of inclusive fitness was broader than the process now known as kin selection.

Following Dawkins, a greenbeard is usually defined as a gene, or set of closely linked genes, that has three effects (Gardner and West 2010; West and Gardner 2010; Madgwick et al. 2019):

1. It gives carriers of the gene a phenotypic label, such as a green beard.
2. The carrier can recognize other individuals with the same label.
3. The carrier then behaves nepotistically towards individuals with the same label.

Dawkins's presentation of Hamilton's insight is striking, but its cartoon-like character obscures some of the general properties of the greenbeard effect. For example, the phenotypic marker (the greenbeard) is not required. The key feature is that the gene for social behaviour is linked to the genetic basis of the assortment mechanism, which can happen if the genes for nepotistic behaviour and habitat preference are linked (Hamilton 1975). For example, if the greenbeard locus makes individuals follow a certain flower scent and then behave nepotistically towards other individuals at that flower (Madgwick et al. 2019).

The greenbeard effect should be distinguished from what Dawkins called the 'armpit effect' (Dawkins 1982, p. 146). The latter refers to situations where an organism recognizes a trait (like odour) in itself, or in a close relative, and then favours other individuals with the same trait. As such, it provides a mechanism for traditional kin-recognition. Greenbeard genes work to recognize copies of themselves, not of kin. This is also why the armpit effect is expected to be mediated by high genome-wide relatedness, whereas the greenbeard effect relies only on high relatedness at the locus/loci underlying it.

5.3.1 Why greenbeards should be rare in nature

The conditions for their existence meant that greenbeards were long thought to be unlikely to exist in nature (Dawkins 1978, 1979; Grafen 1984). Madgwick et al. (2019) group arguments for why greenbeards should be rare in natural populations into three categories: existential, detection, and selection arguments.

The existential argument points out that greenbeards rely on the assumption that a single gene, or a set of tightly linked genes, can affect all three phenotypic functions required. This seems unlikely, especially as the connection of genotype to phenotype will need to be deterministic, or close to deterministic. Combined, these improbable conditions mean that greenbeards should evolve only very rarely (West et al. 2007). In response, Haig has noted that there are several examples of genes, or sets of closely linked genes, that can recognize themselves in others, including mechanisms like cell–cell receptors (Haig 1996, 2013). Furthermore, a greenbeard gene does not need to directly cause the traits; it can achieve its effect by adjusting the expression of other genes. The kinds of phenotypes that can be regulated in this way may be limited, which could explain why many of the best examples of greenbeards come from microbes, whose simpler genetic architectures can prevent pleiotropic traits from being broken up, ensuring that beards and behaviour stay linked.

The detection argument holds that even if greenbeards do evolve, they will be difficult for researchers to notice in nature. The reason for this is that greenbeard genes are expected to evolve under strong selection and so will rapidly spread to fixation in the population (Biernaskie et al. 2013). At that point, the greenbeard would lose its effect. There seem to be biological reasons for why this does not always happen. Take, for example, the *Gp-9* greenbeard gene in the red fire ant (*Solenopsis invicta*), the first example of a greenbeard found in nature (Keller and Ross 1998; Ross and Keller 2002). *Gp-9* is involved in the production of an odour-binding protein and comes in two alleles (*B* and *b*). The recessive allele (*b*) acts as a greenbeard by causing workers to use the odour to detect queens that lack the allele and kill them. Queen-killing can happen because the queens in populations with multiple queens are heterozygous (*Bb*) at the *Gp-9* locus. These *Bb* queens produce *BB* and *Bb* offspring (*bb* offspring die young). *BB* queens would give rise to mongyne colonies (i.e. with only one queen), were they not killed by *Bb* workers. The homozygous lethality means that *b* will not spread to fixation, and the resulting polymorphism is what allows the greenbeard to be detected.

Another reason that greenbeards may be difficult to detect is that they may be in conflict with the rest of the genome (Alexander and Borgia 1978). Such a situation arises when the relatedness coefficient between two individuals at the greenbeard locus is different from the genome-wide average. Whereas it is in the interest of the greenbeard locus to favour another greenbeard individual, this interest may not be shared by other genes in the genome. For some time confusion existed in the literature about the generality of this situation. For example, Ridley and Grafen (1981) demonstrated that a suppressor mutation that prevents the greenbeard from being expressed will generally be favoured only when the greenbeard itself is under negative selection. While suppressing the greenbeard saves the individual the cost of behaving favourably towards other greenbeards, the suppression also prevents the individual from reaping the rewards of receiving altruistic acts from other greenbeards. Ridley and Grafen further argued that if a suppressor could keep the beard while simultaneously suppressing the altruistic act, this false-beard situation would create the stage for conflict. Biernaskie et al. (2011) clarified the situation by developing models of greenbeards and false beards with and without suppressors. They showed that the fitness interests of greenbeards will generally be in line with the rest of the genome. The exception, conflict, occurs when a greenbeard individual is interacting with closely related kin. Here, kin selection will favour altruistic behaviour towards all close relatives, but the greenbeard locus will be selected to restrict it to fellow greenbeards.

Finally, the selection argument is that greenbeards are unlikely to be selected for to begin with. Because the benefit of having a beard is only realized when there are other beards around, several mathematical models have shown that greenbeards may struggle to increase at low frequencies (Jansen and van Baalen 2006; Gardner and West 2010; Biernaskie et al. 2011). While limiting the opportunity for greenbeard genes to invade a population, demographic factors like population structure can readily provide the right circumstances. Although new greenbeard mutations will be rare in the general population, they may be locally common within a kin group.

5.3.2 Helping and harming greenbeards

Despite the expectation that greenbeards should be rare in natural populations, several examples have been identified. One particularly well-researched example is that of clumping in yeast (*Saccharomyces sp.*; reviewed in El-Kirat-Chatel et al. 2015). In response to stress, thousands of yeast cells come together to form large clumps ('flocs') that resemble bacterial biofilms. This cooperative behaviour provides protection from various environmental stressors, including anti-microbial substances and ethanol, as cells on the inside of the floc are physically shielded from the external environment. The protection, however, comes with a fitness cost, which has been demonstrated by growing flocculating and non-flocculating cells under non-stressful conditions. In such a benign environment, non-flocculating cells grow four times faster than flocculating cells. Flocculation has been used for centuries in the brewing industry to remove excess yeast during alcohol production, but it is typically absent in lab strains (Speers 2012).

The genetics of flocculation has also been extensively studied. Smukalla et al. (2008) identified the flocculating gene *FLO1* as a greenbeard locus in *S. cerevisiae*. The *FLO1* locus produces a trans-membrane protein that allows yeast cells to bind to one another. Cells that share the cooperative *FLO1* allele flocculate together, while those that lack the cooperative allele are mostly excluded from flocs. The tractability of the *FLO1* system offers an excellent opportunity to experimentally teach students which aspect of inclusive fitness theory—kin selection or the greenbeard effect—best explains a cooperative behaviour, as thousands of undergraduates at the University of Toronto have come to experience (Ågren et al. 2017).

The *Gp-9* and *FLO-1* examples differ both in the nature of their host organism and in the way that the greenbeard effects comes about. In his original popularization of the concept, Dawkins outlined a scenario where greenbeards help fellow greenbeards (Dawkins 1976, pp. 115–116, 1982a, pp. 143–155). This is what happens in the *FLO-1* example and also when slime mould (*Dictyostelium discoideum*) individuals with the same *csa* allele stick together in fruiting bodies, excluding individuals with other alleles (Queller et al. 2003). In contrast, the queen killing in the red fire ant is an example of a greenbeard effect that is achieved by greenbeards actively harming non-greenbeards. Greenbeards can therefore generally be divided into 'helping' and 'harming' greenbeards (Gardner and West 2010). Additional examples of each kind have been described, including helping greenbeards in the fungus *Neurospora crassa* (Heller et al. 2016) and the marine tunicate *Botryllus schlosseri* (De Tomaso 2018). Bacteria seem to have examples of both helping and harming greenbeards (Riley and Wertz 2002; Pathak et al. 2013).

These empirical examples combine with theoretical advances connecting the evolutionary dynamics of greenbeards to everything from the evolution of warning coloration (Guilford 1985) to sexual selection (Faria et al. 2018), to transform what was once a cute amusing thought experiment into chloropogonology the thriving study of greenbeards (Gardner and West 2010). Relevant to the gene's-eye view, greenbeard genes highlight how the oft-assumed unity of purpose of the genome can break down. The potential for such breakdown is even greater in situations involving selfish genetic elements.

5.4 Selfish genetic elements

Traditionally, the predominant view of genomes was that of a highly coordinated network, with all parts working together for the same purpose. Genetic transmission was assumed to be fair and to be governed by the rules laid out by Mendel. The traditional view is challenged by the existence of stretches of DNA that can subvert these rules and promote their own transmission at the expense of other parts in the genome, with either no or a negative effect on organismal fitness. Such stretches of DNA are now usually called selfish genetic elements, but they have previously been known as parasitic DNA, selfish DNA, ultra-selfish genes, genomic outlaws, and self-promoting elements (reviewed in Werren et al. 1988; Burt and Trivers 2006; Werren 2011; Gardner and Úbeda 2017; Ågren and Clark 2018).

5.4.1 Early work on selfish genetic elements

Though foreshadowed by T.H. Huxley's ideas about competition among gemmules (Huxley 1878), proper discussions of selfish genetic elements began in earnest a few decades later. In 1928, the Russian geneticist Sergey Gershenson reported the discovery of a driving X chromosome that was inherited in more than 50% of the gametes in

Drosophila obscura (Gershenson 1928). Crucially, he noted that the resulting female-biased sex ratio could potentially drive a population extinct. Haldane discussed several examples of conflict between different levels in the biological hierarchy, including how pollen competition could lead to the spread of traits that were deleterious to the individual plant (Haldane 1932, p. 67). In 1945, the Swedish botanist and cytogeneticist Gunnar Östergren argued that supernumerary (i.e. non-vital) B chromosomes were best conceived as parasitic (Östergen 1945). Östergren's paper was the first unambiguous articulation of the idea of selfish genetic elements.

Östergren's work coincided with several similar observations, particularly in plants (Ågren and Wright 2015). For example, female meiotic drive was first reported in maize (Rhoades 1942), and Lewis (1941) presented evidence that cytoplasmic male sterility in plants resulted from the conflict between maternally inherited organellar and biparentally inherited nuclear genes. Then, in the early 1950s, Barbara McClintock published a series of papers describing the existence of transposable elements, which are now recognized to be among the most successful selfish genetic elements there are (McClintock 1950, 1956; though McClintock herself rejected the selfish label). Her discovery of transposable elements led to the Nobel Prize in Medicine or Physiology in 1983 and she remains the only woman to have been awarded the prize on her own. These examples were initially considered to be genetic oddities with few implications for evolutionary theory (Burt and Trivers 2006, pp. 12–16; Werren 2011). It would take several decades before selfish genetic elements in general, and their evolutionary implications in particular, became widely appreciated.

The gene's-eye view played an important role in bringing selfish genetic elements to the high table of genetics and evolutionary biology. Viewing evolution as a struggle between competing replicators made it easier to recognize that not all genes in an organism would share the same evolutionary fate (Rothstein and Barash 1983). As Dawkins put it:

> It is a remarkable fact that natural selection seems to have chosen those replicators that cooperate with each other and go around in the large collective packages which we see as individual organisms. This is a fact that needs explaining in its own right, just as the existence of sexual reproduction needs explaining in its own right.   (Dawkins 1978)

In addition to the general conceptual shift to gene-level thinking in evolutionary biology, empirical observations also led the way in the origin of the study of selfish genetic elements (Werren 2011). Early work on genome structure reported that large chunks of eukaryotic genomes were made up of genetic material, such as repetitive DNA, with seemingly no connection to organismal function or fitness (see, e.g., Britten and Kohne 1968). Moreover, while it was clear that genome size varies dramatically across species (it is now known that eukaryotes vary more than 60,000-fold; Elliott and Gregory 2015), there was no correlation between the amount of DNA in an organism and its perceived complexity. For example, the genome of a single-celled amoeba is about 100 times larger than that of humans.

These accumulating empirical observations were central to two papers published back-to-back in *Nature* in 1980, by Leslie Orgel and Francis Crick, and Ford Doolittle and Carmen Sapienza respectively. Both papers attempted to counter the prevailing view of the time that the presence of differential amounts of non-coding DNA and transposable elements was best explained from the perspective of individual fitness, described as the 'phenotypic paradigm' by Doolittle and Sapienza. The authors argued that much of the genetic material in eukaryotic genomes persists, not because of its phenotypic effects, but, citing Dawkins, that it can be understood from a gene's-eye view without invoking individual-level explanations. These papers resulted in a series of exchanges in *Nature* (Dover 1980; Cavalier-Smith 1980; Orgel et al. 1980; Dover and Doolittle 1980; Jain 1980) representing the first high profile discussion of the evolutionary implications of selfish genetic elements.

These papers marked the beginning of the serious study of selfish genetic elements, and the subsequent decades saw a rapid increase in both theoretical advances and empirical discoveries. Leda Cosmides and John Tooby wrote a landmark review about the conflict between maternally inherited organellar genes and biparentally inherited nuclear genes (Cosmides and Tooby 1981; see also Eberhard 1980). Their paper used the gene's-eye view to provide a comprehensive introduction to the logic of genomic conflicts, foreshadowing many themes that would later become research foci. In 1988, Jack Werren and colleagues wrote the first major empirical review of the topic (Werren et al. 1988). In it, they coined the term selfish genetic element, putting an end to the confusingly diverse terminology mentioned above. It was also the first paper to bring together all the different kinds of selfish genetic elements known at the time, discussing examples ranging from meiotic drive and supernumerary B chromosomes to killer plasmids, selfish mitochondria, and transposable elements.

In the 1980s, selfish genetic elements seem to have been considered an exception of limited interest. By 2006, when Austin Burt and Robert Trivers published the first book-length treatment of the topic, a comprehensive piece that remains the go-to source, the tide was changing. While the role of selfish genetic elements in evolution long remained controversial, a recent review concluded with a statement that no longer feels particularly radical: 'nothing in genetics makes sense except in the light of genomic conflicts' (Rice 2013).

5.4.2 Examples of selfish genetic elements

Selfish genetic elements have now been described in most groups of organisms, and they demonstrate a remarkable diversity in the ways they promote their own transmission. Below, I discuss some examples of this diversity.

Genomic conflicts often arise because not all genes are inherited in the same way. Probably the best example of this is the conflict between uniparentally (usually, but not always, maternally) inherited mitochondrial and biparentally inherited nuclear genes (Havird et al.

2019; Ågren et al. 2019b). The conflict between mitochondrial and nuclear genes is especially well-studied in flowering plants. Flowering plants are typically hermaphrodites and mitochondrial genes are usually only transmitted through female gametes. From their point of view, the production of pollen leads to an evolutionary dead end. Any mitochondrial mutation that can affect the amount of resources the hermaphroditic plant invests in the female reproductive functions at the expense of the male reproductive functions improves its own chance of transmission. Cytoplasmic male sterility, then, is the loss of male fertility, typically through loss of functional pollen production, resulting from a mitochondrial mutation (Case et al. 2016). In many species where cytoplasmic male sterility occurs, the nuclear genome has evolved so-called restorer genes, which repress the effects of the cytoplasmic male sterility genes and restore the male function, making the plant a hermaphrodite again.

Mito-nuclear conflict in plants is a good example of the strength and limitations of the gene's-eye view. It provides a clear logic as to why transmission asymmetries should result in genomic conflicts. At the same time, it has little to offer on the question why there are no reported examples of conflict between nuclear and chloroplast genes, which like mitochondrial genes are usually uniparentally inherited.

A related consequence of the maternal inheritance of the mitochondrial genome is the so-called Mother's Curse. This is the name given to the fact that because genes in the mitochondrial genome are strictly maternally inherited, mutations that are beneficial in females can spread in a population even if they are deleterious in males (Gemmell et al. 2004). A key observation consistent with this idea is that mitochondrial genetic disease in humans seem to affect males more than females (Frank and Hurst 1996). Explicit screens in fruit flies have also successfully identified female-neutral but male-harming mitochondrial mutations (Camus et al. 2012; Patel et al. 2017). In humans, a 2017 paper studying the prevalence of Leber's hereditary optic neuropathy, an eye disease, in Quebec, Canada, showed how the mutation causing the disease is present at an unusually high frequency given the severity of the syndrome. The way to make sense of this is that the disease-causing mutation is located in the mitochondria and that the disease affects males more than females (Milot et al. 2017). In an extra twist to the story, the authors were able to demonstrate that the mutation was introduced to the Quebec population by one of the so-called *Filles du roi* (King's Daughters) who were sent there in the 17th century by the French King Louis XIV in an attempt to help populate the colony of New France.

Another way that selfish genetic elements can manipulate the transmission to their own advantage is to interfere directly with the process of meiosis. Because sexual reproduction results in a zygote that is the product of the mixing of genes from two individuals, an opportunity for competition between nuclear genes in each parent arises over who makes it into the zygote. This potential conflict is mainly avoided by the imposition of fair meiosis (Leigh 1971; Haig and Grafen 1991; Frank 2003). Still, there are several ways genes can cause unfair meiosis and end up being overrepresented in gametes (Lindholm et al. 2016). One example from female meiosis involves genes that ensure that they are preferentially transmitted to the egg cell, as opposed to one of the two polar bodies. Because polar bodies are not fertilized, genes with this ability are guaranteed to be transmitted to the next generation. Another example are so-called sperm killers, where genes damage the development of sperms in which they are absent. Two of the best studied examples of this selfish strategy are the *Segregation Distorter* in *Drosophila melanogaster* (Larracuente and Presgraves 2012) and the *t*-haplotype of the domestic mouse *Mus musculus* (Herrmann and Bauer 2012).

A closely related example is B chromosomes. These are chromosomes that are not required for the viability or fertility of the organism but exist in addition to the normal set (the so-called A set; Jones 1991; Douglas and Birchler 2017; Benetta et al. 2019). B chromosomes persist in the population and accumulate because they have the ability to propagate themselves faster than the A chromosomes. Though typically smaller than other chromosomes, their gene poor, heterochromatin-rich structure made them visible to early cytogenetic techniques and B chromosomes were first detected over a century ago (Wilson 1907; see also Östergren 1945). They have been thoroughly studied and are estimated to occur in 15% of all eukaryotic species (Beukeboom 1994) and appear to be particularly common among eudicot plants, while rare in birds and mammals. The phenotypic consequences of B chromosomes are unclear, but copy number correlates positively with genome size (Trivers et al. 2004) and has also been linked to a decrease in egg production in grasshoppers (Zurita et al. 1998).

A kind of selfish genetic element that has received a lot of attention in recent years are homing endonucleases. These are enzymes that cut DNA in a sequence-specific way, and those cuts are then 'healed' by the regular DNA repair machinery (Burt et al. 2004; Belfort and Bonocora 2014). Because homing endonucleases cause these cuts at the site homologous to the first insertion site, this results in a conversion of a heterozygote into a homozygote. CRISPR-Cas9 technology allows the artificial construction of homing endonuclease systems, so-called 'gene drive' systems. Gene drives have the potential to introduce a desired allele in a population, with the ability, for example, to exterminate malaria-carrying mosquitos, and so present a combination of great promise for biocontrol and risk to ecosystem dynamics (see discussions by, e.g., Esvelt et al. 2014 and Champer et al. 2019).

The most successful kind of selfish genetic element, at least in terms of abundance, are transposable elements. This group includes a wide variety of DNA sequences that all have the ability to move to new locations in the genome of their host. DNA transposons do this by a direct cut-and-paste mechanism, whereas retrotransposons need to produce an RNA intermediate to move, analogous to a copy-and-paste mechanism. Regardless of the specific mechanism of self-replication, most transposon insertions appear to be relatively innocuous (Lisch 2013). For example, the colour difference between the wine grapes of Cabernet, Chardonnay, and Ruby Okuyama are due to the insertion (Cabernet to Chardonnay) and subsequent loss

(Chardonnay to Ruby Okuyama) of a specific retrotransposon. Genomes also seem fairly tolerant of the presence of transposons in their genomes. A sizable portion of the genome of many animals and plants is made up of transposons, including some 50% of the human genome and more than 80% of that of maize (Tenaillon et al. 2010; Kapusta et al. 2017).

The movement of transposons is also a source of mutations, some with devastating results. For example, the presence of transposons has been linked to human diseases ranging from cancer to haemophilia (Hancks and Kazazian 2016). Both plants and animals have therefore evolved means for reducing the deleterious fitness consequences of transposable elements, often through quite intricate small RNA interference mechanisms (Blumenstiel 2011; Kelleher et al. 2020). Because new insertions can disrupt gene function, sometimes those changes can have positive fitness effects, just like any other kind of mutation. Examples of transposon-driven adaptive changes range from *Drosophila* (Aminetzach et al. 2005) to dogs (Cordaux and Batzer 2006), but the most charismatic example is probably the discovery that the mutation underlying the industrial melanism in the peppered moth is a transposable element insertion (van't Hof et al. 2016).

5.4.3 Two rules for selfish genetic elements

Though selfish genetic elements demonstrate a tremendous diversity in how they alter the rules of transmission to promote their own interests, some commonalities can be identified. In a 2001 review, Greg Hurst and Jack Werren proposed two such 'rules' for selfish genetic elements (Hurst and Werren 2001). Both demonstrate the power of the gene-centred thinking.

Their first rule is that selfish genetic elements need sex to spread. In asexual lineages, selfish genetic element are essentially stuck, as the same genome is passed on intact from parent to offspring. This will increase the variation in fitness among individuals, resulting in stronger purifying selection in asexually reproducing lineages, for a lineage without the selfish genetic elements should outcompete a lineage with the selfish genetic element. Sex, in contrast, allows selfish genetic elements to spread into new genetic lineages. In line with this, population genetic theory generally predicts the absence of sex is expected to lead to a reduction in the number of selfish genetic elements present in the genome (Hickey 1982; Wright and Schoen 1999; Dolgin and Charlesworth 2006).

An important caveat to this prediction is that the efficacy of selection is expected to be reduced in asexual lineages due to the low rates of recombination. Such a reduction may result in the accumulation of deleterious mutations, which includes most selfish genetic elements, in a process known as a Muller's ratchet (reviewed in Hartfield 2015). Mathematical modelling has shown that asexual lineages may be driven to extinction through an indefinite accumulation of transposable elements (Hickey 1982; Dolgin and Charlesworth 2006). Many of the predictions about the abundance of selfish genetic elements in sexual versus asexual lineages also apply when comparing self-fertilizing and outcrossing lineages.

To date, the empirical evidence for the importance of sex and outcrossing comes from a variety of selfish genetic elements. Examples include finding a higher abundance of self-promoting plasmids in sexual compared to asexual yeast strains (Harrison et al. 2014) and more B chromosomes in outcrossing than self-fertilizing plants (Burt and Trivers 1998). For transposable elements, the results are mixed. Studies in plants (Ågren et al. 2014), animals (Bast et al. 2016; Szitenberg et al. 2016), and fungi (Bast et al. 2019) have reported differences in both directions and neither.

The second rule is that hybrids are often required to reveal the presence of selfish genetic elements in a population. There are two key reasons for this. First, selfish genetic elements can rapidly sweep to fixation. Unless a specific element is segregating in the population, so that some individuals carry it and others do not, its presence is very difficult to detect. Hybridization events, however, may produce offspring with and without the specific selfish genetic elements and so reveal their presence. Second, host genomes have evolved mechanisms to suppress the activity of the selfish genetic elements. Examples include the silencing of transposable elements by small RNAs and nuclear genes that restore male fertility in the face of selfish mitochondria. If the co-evolution of selfish genetic elements and their suppressors is rapid, that can also mask the presence of selfish genetic elements in a population. Hybrid offspring, on the other hand, may inherit a selfish genetic element but not its corresponding suppressor and so reveal its presence.

Co-evolution between selfish genetic elements and their suppressors can also cause reproductive isolation through so-called Bateson–Dobzhansky–Muller incompatibilities (Crespi and Nosil 2013; Ågren 2013; Serrato-Capuchina and Matute 2018). An early striking example of hybrid dysgenesis induced by a selfish genetic element is the *P* element in *Drosophila melanogaster* (Kidwell 1983). When males carrying the *P* element were crossed to females lacking it, the resulting offspring suffered from reduced fitness, whereas offspring of the reciprocal cross were normal (this is expected because piRNAs are maternally inherited). The *P* element is typically present only in wild strains, not in the lab strains of *Drosophila melanogaster* that were collected before the *P* elements became widespread in the species. Recently the *P* element has invaded natural populations of the closely related *D. simulans,* where it is also causing hybrid dysgenesis (Kofler et al. 2015; Hill et al. 2016).

The *P* element story is a good example of how the rapid co-evolution of selfish genetic elements with their silencers can quickly lead to incompatibilities in as little as a few decades. Several other examples of selfish genetic elements causing reproductive isolation have since been demonstrated. Hybrid dysgenesis has been shown to be associated with centromeric drive in barley (Sanei et al. 2011), sex ratio distorters in flies (Verspoor et al. 2018), and in several species of flowering plants by mito–nuclear conflict (Rieseberg and Blackman 2010). After much initial skepticism regarding the potential role of selfish genetic elements in speciation (reviewed in Patten 2018), attitudes in the field seem to be changing.

5.4.4 Selfish genetic elements and selfish genes

In an interview with Ullica Segerstrale, Robert Trivers said 'I will get Lewontin at the molecular level!' (Segerstrale 2000, p. 330). In addition to being a long-term critic of the gene's-eye view, Lewontin had also been at the forefront of bringing molecular data into evolutionary analysis (e.g. Hubby and Lewontin 1966; Lewontin and Hubby 1966). Trivers, in contrast, had been instrumental in the emergence of the gene's-eye view and was one of the first to see the conceptual complementarity of selfish genetic elements and selfish genes in the more general Dawkinsian sense. Together with Austin Burt, with whom he began writing *Genes in Conflict* in the early 1990s (Burt and Trivers 2006, p. vii), he had a particularly good understanding of the significance of the molecular data on selfish genetic elements. To Trivers, the abundance of data on selfish genetic elements would provide the ultimate vindication of the superiority of the gene's-eye view.

The relationship between the gene's-eye view and the study of selfish genetic elements may in retrospect seem straightforward. Just consider the similarity in language in two key papers in each field: Hamilton's first paper, the 1963 note in the *American Naturalist*, and Östergren's 1945 paper on B chromosomes mentioned above:

> Despite the principle of 'survival of the fittest' the ultimate criterion that determines whether [a gene for altruism] G will spread is not whether the behavior is to the benefit of the behaver but whether it is of benefit to the gene G.   (Hamilton 1963)
>
>   In many cases these chromosomes have no useful function at all to the species carrying them, but that they often lead an exclusively parasitic existence...[B chromosomes] need not be useful for the plants. They need only be useful to themselves.   (Östergren 1945)

A crucial conceptual insight in both papers is that in order to explain the phenomenon under study, the origin of altruism and the spread of B-chromosomes respectively, the investigator is better off viewing the world from the perspective of genes rather than individual organisms.

While the evolutionary logic of altruism and selfish genetic elements is difficult to follow from an organismal perspective, it is straightforward from a gene's-eye view.

The existence of selfish genetic elements has often been seen as one of the strongest arguments for the approach (Okasha 2006, p. 153, 2012; Ågren 2016; Ridley 2016; De Tiège et al. 2018). For example, when discussing the work of Eberhard (1980) and Cosmides and Tooby (1981) on the conflicts between the nuclear and organellar genomes, Dawkins (1982a) wrote: 'Neither Eberhart nor Cosmides and Tooby explicitly justify or document the genes'-eye view of life: they simply assume it (...) These papers have what I can only describe as the flavour of post-revolutionary normal science' (Dawkins 1982a, p. 178). This is not the whole story, however. In *Adaptation and Natural Selection*, Williams discussed only one example of a selfish genetic element, the *t*-haplotype in mice studied by Lewontin (Lewontin and Dunn 1960; Lewontin 1962). Ironically, the inability of the *t*-haplotype to spread to high frequencies is presented by Williams (1996, p. 117), as well as by Leigh (1971, p. 247), as the only convincing case of group selection in nature.

The tension between selfish genetic elements and other levels of selection was also central to Gould's argument for a hierarchical approach to evolutionary theory. In *The Structure of Evolutionary Theory*, he wrote:

> When future historians chronicle the interesting failure of exclusive gene selectionism (based largely on the confusion of bookkeeping with causality), and the growing acceptance of an opposite hierarchical model, I predict that they will identify a central irony in the embrace by gene selectionists of a special class of data [selfish genetic elements], mistakenly read as crucial support, but actually providing strong evidence of their central error.   (Gould 2002, p. 689)

How to make sense of Gould's critique? One way is to recall the discussion of pluralistic gene selectionism in Chapter 3 and note several versions of the gene's-eye view can be said to exist. The strongest version argues that the gene level is the only level of selection, and talking about other levels is misleading. This is how much of *The Selfish Gene* is written. Other, weaker, versions emphasize that the Necker cube approach to genic and individual selection and/or suggest that all instances of evolution can at least be represented as changes in allele frequency. The least controversial version simply states that in some cases, for example selfish genetic elements, the gene is the level of selection.

Above Gould is attacking the strongest version of the gene's-eye view, the argument that only the gene level perspective truly represents evolution by natural selection. Although he never warmed to the term 'selfish DNA' (as selfish genetic elements were often called during Gould's active years), as he thought it privileged the individual organism in an inappropriate way, he did consider selfish genetic elements to be strong evidence for a hierarchical view of evolutionary biology (Gould 1977, 1983a). The link between within-genome selection and hierarchy was later been picked up and expanded by several others (Vrba and Eldredge 1984; Doolittle 1989; Gregory 2004, 2013; Gregory et al. 2016). The main argument of these authors is that evolutionary explanations of selfish genetic elements must involve selection at both the level of the selfish genetic element and at the level of the individual organism. Like proponents of a gene's-eye view, Gould and his intellectual allies strive to demote the individual from a central position in evolutionary theory. The two groups part ways in that Gould and friends want to add additional levels to the hierarchy, whereas gene selectionists insist on focusing on one level, that of the gene.

Given Gould's aversion for adaptationism, it is surprising that he did not pick up on how the gene's-eye view and the study of selfish genetic elements can act to counter naive adaptationist thinking. Instead, this message was deliver by two representatives of evolutionary psychology, often considered the worst offenders of uncritical adaptationism, Leda Cosmides and John Tooby. The pair made the point in a sharply worded letter to the editor of *The New York Review of Books* responding to two of Gould's essays that touched on evolutionary psychology (Gould 1997a, 1997b; the letter was rejected but uploaded online, Tooby and Cosmides 1997). Together with Jereme Barkow, Cosmides and Tooby were the editors of *The Adapted Mind: Evolutionary Psychology and the Generation of Culture* (Barkow et al. 1992), often considered the foundational text of the field. As noted above, Cosmides and Tooby had also made a substantial contribution to the study of genomic conflict with their 1981 paper (Cosmides and Tooby 1981). A key point of their letter was that a gene's-eye view forces biologists to look beyond individual organisms to see the amount of maladaptation caused by selfish genetic elements. Since that letter, the appreciation of the phenotypic consequences, beneficial and deleterious and neutral, of selfish genetic elements has increased (Lisch 2013; Rice 2013), and it has become more difficult to ignore the existence of competing genes within individuals when the goal is to develop a general account of adaptation.

As discussed in Chapter 4, the idea that individual organisms act to maximize their inclusive fitness is based on the assumption of genomic unity of purpose. Selfish genetic elements show that not all genes share the same fitness interests and rather than being a cohesive fitness maximizer, the individual organism is a compromise of several fitness interests. A recent paper set out to tackle this contradiction (Scott and West 2019). They develop a series of mathematical models of selfish genetic elements and suppressors to argue that selfish genetic elements will either impose fitness costs that are too weak to be important, or, if the fitness costs to the individual organism are large, they will be readily suppressed. From this, they argue that even if selfish genetic elements are common, their effects will be negligible and can be ignored, thus maintaining the idea of individual level fitness maximization.

There are issues with this conclusion. As Scott (2019, p. 211) points out there is an implicit bias in their approach that saves the individual. They assume that the individual is already maximizing its fitness, which results in a situation where a selfish genetic element cannot push the individual away from that maximization. What they show is not how fitness maximization can be achieved in the presence of selfish genetic elements in the first place, but how it can be maintained once it is there.

5.4.5 The only game in town?

Just as trying to explain everything in terms of individual fitness makes biologists ignore too much interesting biology, too much focus on individual genes can also lead one astray. For example, those seduced by the gene's-eye view can overemphasize genomic conflict at the expense of genomic cooperation (Wade and Drown 2016). How genes cooperate has received much less attention than how whole organism cooperate (Ågren 2014; Yanai and Lercher 2016). A noticeable exception is research on cooperation in the RNA world (Higgs and Lehman 2015), which is a potential empirical goldmine to be approached with the tools of the gene's-eye view.

The gene's-eye view is arguably the most powerful way to think about selfish genetic elements. It is, however, not the only way. Additional frameworks highlight other aspects of their biology. To date, conceptual and empirical insights have come from several traditions including, but not limited to, multi-level selection theory (Gregory 2013; Gregory et al. 2016), host–parasite interactions (Nee and Maynard Smith 1990; Hurst 1996; Brookfield 2011), political philosophy (Okasha 2012), epidemiology (Wagner 2009), and community ecology (Venner et al. 2009; Linquist et al. 2015). All frameworks are not created equal, but most can contribute something. Studying of selfish genetic elements is difficult; we need all the help we can get.

5.5 Summary

- The emergence of the gene's-eye view led or contributed to the development of several new empirical research avenues. Chief among these were the study of extended phenotypes, greenbeards, and selfish genetic elements.

- Extended phenotypes include 'all the effects that a gene causes on the world' and can be grouped into three main categories: animal architecture, action at a distance, and parasite manipulation of host behaviour.

- Greenbeard genes are genes that can identify copies of themselves in other individuals and then make their carrier act nepotiscally towards those individuals or selfishly towards non-carriers. The term comes from a thought experiment meant to demonstrate that from a gene's perspective, what matters is the relatedness at a particular locus that causes the social act, not the average relatedness across the genome. Although initially the conditions for their existence were thought to be too restrictive, many examples of greenbeard genes have been identified in natural populations.

- Selfish genetic elements have the ability to promote their own transmission despite having a harmful effect on the fitness of the individual organism. They come in a remarkable diversity of flavours and they are a prominent feature of nearly all sexually reproducing organisms.

- In their own ways, extended phenotypes, greenbeards, and selfish genetic elements, all highlight the value of thinking from a gene's perspective, and strengthen the gene's-eye view's attempt to dethrone the individual organism as the central unit of explanation in evolutionary biology.

Conclusion: The Gene's-Eye View Today

Marian Stamp Dawkins once recounted her experience marking a final exam paper by one of her undergraduates (Stamp Dawkins 2006). The student successfully answered the set question on how natural selection could have resulted in some animals not reproducing, using all the correct technical terms, but then went on to say 'And here I rely heavily on the words of Richard Dawkins' before wrapping up their conclusion. In response, Stamp Dawkins noted in the margin of the examination paper: 'Yes. Don't we all?'

There is no denying that for the past decades the field of evolutionary biology—students and faculty alike—have relied much on the words of Richard Dawkins. As this book has shown, people disagree whether this is a good thing or not.

In *The Descent of Man*, Darwin distinguished between 'false facts' and 'false views' (Darwin 1871, p. 385). False facts, he argued, were directly detrimental to the progress of science as they had a tendency to linger on. In contrast, false views that had some empirical support, could actually be beneficial to a field, as 'every one takes a salutary pleasure in proving their falseness'. Always one for coining a catchy phrase, Gould christened these kinds of views that were incorrect, but nevertheless fruitful, Pareto errors, after the economist Vilfredo Pareto who is meant to have said: 'give me the fruitful error any time, full of seeds, bursting with its own corrections. You can keep your sterile truth for yourself' (quoted in Gould 2002, p. 612).

The Gene's-Eye View of Evolution. J. Arvid Ågren, Oxford University Press. © J. Arvid Ågren 2021.
DOI: 10.1093/oso/9780198862260.003.0007

To Gould, there was no bigger Pareto error in evolutionary biology than the gene's-eye view. This book is my answer to Gould's accusation. The gene's-eye view is one way to think about biology, one that works especially well when we want to work out the logic of evolutionary scenarios. It achieves this success by being agnostic about many biological details, including what exactly a gene is or a how development actually works. As a consequence, it loses potency in situations where those details matter.

So far, the book has traced the origin and development of the gene's-eye view, clarified typical points of contention, and shown how it helped us think about old evolutionary problems and pointed out new ones. Biology has changed a lot in the half century since *Adaptation and Natural Selection* and *The Selfish Gene* were written. In these last few pages, I will look ahead and offer some reflections on the standing of the gene's-eye view and the nature of the field of evolutionary biology.

Why should biologists study the history of ideas?

A computational biology colleague of mine once described evolutionary biology as a field obsessed with its own history. That might well be true, but if so the obsession goes back a long way. Take, for example, R.A. Fisher who wrote:

> More attention to the History of Science is needed, as much by scientists as by historians, and especially by biologists, and this should mean a deliberate attempt to understand the thoughts of the great masters of the past, to see in what circumstances or intellectual milieu their ideas were formed, where they took the wrong turning or stopped short on the right track.   (Fisher 1959)

Scientists writing their own history is a double-edged sword. On the one hand, they were there and therefore may know the ideas and people involved. On the other hand, they were there, which means they have vested interests in how these ideas and people fare when history is written. Someone who was straddling that tricky balance was Ernst Mayr, who has had an enormous influence on the way in which we evolutionary biologists view our field, its history, and its philosophy. Robert Trivers once described him as having 'the strongest phenotype of any organism [he had] ever met, man or beast' (Trivers 2009) and books like *The Growth of Biological Thought* (Mayr 1982), *One Long Argument* (Mayr 1991), and *This is Biology* (Mayr 1998) have been read by many aspiring evolutionary biologists. Readers of Mayr's books, however, receive one version of the history of biology. In particular, it quickly becomes clear that the ideas that Mayr himself held dear (most notable the biological species concept) fare rather well in his version of events.

This book is not a history book, but history plays an important role and so does philosophy. I, like Mayr, am a biologist by training and temperament, not a historian or a philosopher of science, and the book should be read with that in mind.

Another issue with scientists writing their own history is the question of who the intended audience is. Here, Mayr's Harvard colleague Stephen Jay Gould serves as a good example. Gould featured extensively in some chapters of this book, as one of the most articulate and public critics of the gene's-eye view. In many ways, he filled the role of critic-in-chief admirably (see, e.g., Dawkins's tribute 'Unfinished correspondence with a Darwinian heavy weight' published following Gould's death; Dawkins 2003, pp. 218–222). Gould was a fierce opponent of what he considered the 'hardening' of the modern synthesis with its increased focus on selection and gene-level explanations, as well as a tireless advocate of hierarchy, constraints, punctuated rates of change, and non-adaptive evolution (see, e.g., Gould 1980, 1983b; Brown 1999 and Sterelny 2007 provide readable overviews).

Gould often published his criticisms in popular magazines, rather than scientific journals. Such public disagreements over the state of contemporary evolutionary biology put his professional colleagues in an awkward spot. For example, in his review of Daniel Dennett's *Darwin's Dangerous Idea* (Dennett 1995)—a book that was very critical of Gould—John Maynard Smith wrote:

Gould occupies a rather curious position, particularly on his side of the Atlantic. Because of the excellence of his essays, he has come to be seen by non-biologists as the preeminent evolutionary theorist. In contrast, the evolutionary biologists with whom I have discussed his work tend to see him as a man whose ideas are so confused as to be hardly worth bothering with, but as one who should not be publicly criticized because he is at least on our side against the creationists. All this would not matter, were it not that he is giving non-biologists a largely false picture of the state of evolutionary theory.

(Maynard Smith 1995).

Michael Ruse summarized Gould's strategy as a 'rather dangerous game' (Ruse 2013). In his attempt to be taken seriously by his colleagues he was willing to air his grievances in public, in effect forcing professional scientists to listen to him.

Dawkins also largely presented his arguments in public venues and rejected the clear-cut separation between academic and popular writing. But if Gould sold radicalness, Williams and Dawkins offered orthodoxy. This difference in approach is partly why I believe the latter were more successful in their argument. I once got the advice that if you want to change something within a conservative institution (e.g. a university department), the best thing you can do is to present your reform plan as the natural extension of the way things have always been done. If you can successfully frame your proposal in such a way, you can get away with really quite radical change. When Gould put a lot of effort into presenting his ideas as novel and revolutionary—his friend and colleague Richard Lewontin described him as being 'preoccupied with the desire to be considered a very original and great evolutionary theorist' (quoted in Wilson 2015b)—Williams and Dawkins would repeatedly tone down the radical nature of their claims. In *Adaptation and Natural Selection* Williams noted that 'genic selection should be assumed to imply the current conception of natural selection often termed neo-Darwinism' (Williams 1966, p. 96) and Dawkins insisted that the gene's-eye view was simply the modern expression of 'orthodox neo-Darwinism' (e.g. Dawkins 2013a, p. 264).

The gene's-eye view was more radical than that. But through their appeal to orthodoxy Williams and Dawkins were able to introduce comprehensive reform and dethrone the individual organism from the centre of evolutionary explanations and install the gene in its place.

Thinking about the role of individual scientists in history also raises the question of what to do when our scientific heroes disappoint us. I would not have written this book if I did not think Dawkins had made a foundational contribution to our theoretical understanding of evolution. On top of his technical work early in his career, he spent most of his working life introducing evolutionary biology to beginners, as a lecturer at the University of Oxford and to a large global audience through his numerous best-selling books. Lately, he has ruffled many feathers through his advocacy of atheism and he has also been embroiled in various befuddling controversies surrounding race and gender, typically originating on Twitter. In a long, and overall very favourable, article in *The Guardian* titled 'Is Richard Dawkins destroying his reputation?' his friend and intellectual ally Daniel Dennett is quoted worrying that Dawkins was 'seriously damaging his long-term legacy' (Elmhirst 2015). I hope Dawkins's legacy will be the gene's-eye view, but that is a subject for future historians of biology.

For other thinkers featured in this book, it is possible to tell already. I have extensively discussed the ideas of R.A. Fisher, including those first put forward in papers like 'Some hopes of a eugenicist' (Fisher 1914). Recent years have seen increased scrutiny of Fisher's philosophical views, which have prompted several institutions to re-evaluate their affiliation with him. In the span of a few weeks in the spring of 2020, The Society for the Study of Evolution decided to remove his name from the annual award given to the best paper published by a graduate student in the journal *Evolution* and his former Cambridge college debated taking down the stained window commemorating him from its dining hall. Fisher's scientific work guarantees his place as one of the giants of modern biology. That does not mean that we need to celebrate him unquestionably.

Metaphors and mathematics

John Maynard Smith concluded his review of *The Extended Phenotype* with the reflection that neither that book nor *The Selfish Gene* contains a single line of mathematics. Yet, the logic was impeccable (Maynard Smith 1982a). The same compliment can very much be paid to *Adaptation and Natural Selection*: a deeply theoretical argument expressed completely verbally and without equations.

The relationship between biology and mathematics has often been rather awkward. The lack of universal laws that could be captured mathematically made the philosopher J.J.C. Smart deny that biology counted as a proper science in his classic paper 'Can biology be an exact science?' (Smart 1959; see Otsuka 2019). At the same time, mathematics has long played an important role in evolutionary biology. Population genetics has provided the formal backbone since the 1930s, quantitative genetics allows selection to be measured in the lab and in the field, and game theory unified the study of social behaviour in microbes and humans.

The awkward relationship between the fields arises partly because of the kind of mathematical models used in biology. As J.B.S. Haldane put it in his reply to Ernst Mayr beanbag genetics criticism: 'Our mathematics may impress zoologists but do not greatly impress mathematicians' (Haldane 1964). Today the situation is very different. Even if one does not want to go so far as Markowetz (2017) and claim that 'all biology is computational biology', a casual browsing through the major journals in the field reveal that biology can hold its own when it comes to mathematical sophistication. Mathematical models are now an indispensable tool in evolutionary biology (Nowak 2006; Hartl and Clark 2006; Otto and Day 2007). These models come in a variety of shapes and forms. They may involve formal analytical methods, providing exact solutions to different population genetic scenarios. They may be statistical in nature, capturing variances and co-variances of biological traits and fitness. And, when things get really messy, these techniques can be incorporated into computer simulations allowing even more complex problems to be solved. Together, quantitative approaches help us sharpen our intuitions, uncover hidden assumptions, and develop predictions that can be tested empirically.

But there is more to theory in evolutionary biology than mathematical models (Pigliucci 2013; Otsuka 2019). The history of the gene's-eye view shows that verbal-conceptual models are not just a second rate alternative to mathematic models, but the two often actively work in tandem (Servedio et al. 2014; Yanai and Lercher 2020). Verbal-conceptual models often lead the way to new areas that can then be captured in a more formal way.

At the core of many verbal-conceptual models are metaphors. This is true of *The Selfish Gene*, as well as *The Origin of Species* (less so *Adaptation and Natural Selection*). Metaphors are more than simple verbal descriptions (Olson et al. 2019; Kampourakis 2020). Their expressiveness triggers our imagination and stimulates our thinking; they let us see connections that otherwise would have remained obscure. Furthermore, the fact that many metaphors are rather vague is not necessarily the disadvantage it is often portrayed to be, but instead it allows diverse biological phenomena to be brought together.

Like with anthropomorphizing, the use of metaphors tend to make certain scientists uncomfortable (see, e.g., Pauwels 2013). As the pioneering cyberneticists Arturo Rosenblueth and Norbert Wiener warned: 'the price of metaphor is eternal vigilance' (quoted in Lewontin 2001). Good scientific metaphors persist because they help make sense of the world; they make us ask the right questions that can be answered empirically. The best metaphors, like good mathematical models, highlight certain aspects of a biological phenomenon. The flip side is that it ignores other aspects. For instance, the idea of the 'tree of life' beautifully captures the idea of common ancestry and degrees of relatedness. It may also make us resist evidence of hybridization, horizontal gene transfer, and endosymbiosis. Thinking about the evolutionary fate of competing individual genes is a forceful way to sort out the logic of different selective scenarios and the gene's-eye view effortlessly highlights the potential for genomic conflicts. It can also obstruct our focus on genetic drift and developmental constraints. Rosenblueth and Wiener's call for vigilance reminds us that we should always worry not only about what questions a metaphor makes us ask, but also what questions that go unasked.

Alfred Russel Wallace was very concerned with the phrase 'natural selection', which he thought implied a selector. In a 1866 letter, he even tried to convince Darwin to abandon the term in favour of Herbert Spencer's 'survival of the fittest' (Darwin Correspondence Project 2020e). In response, Darwin argued that the worries about the metaphor of natural selection would go away by acquiring 'a little familiarity' with it (Darwin 1868, p. 6). As I set out in the Introduction, I believe the same can be achieved with the gene's-eye view. A little bit of familiarity may also help avoid the issue known as reification. This fallacy occurs when metaphors or abstract constructions are treated as if they were real physical things: 'a map is not the territory' as Korzybski (1933, p. 58) put it. Throughout this book, I have described several instances where critics of the gene's-eye commit this fallacy. At the same time, I have noted how proponents of the perspective, most notably Dawkins, must share some of the blame for this by conflating genic selection as a process and as a perspective.

In general, biology will suffer if there is too much focus on abstract principles. This was the worry of Dawkins's doctoral advisor Niko Tinbergen who late in his career commented on the modern study of behaviour:

> The more abstract thinking of Maynard Smith, Hamilton, Trivers, and Dawkins I do not completely understand, and moreover, I suspect that they know too few animals, and know too little about the multitude of phenomena and aspects.   (Tinbergen 1980, quoted in Kruuk 2003)

Biologists should never lose sight of animals (or other kinds of organisms, from Archaea to orchids). The gene's-eye view is not the only way to think about evolution and natural selection, and for some questions it is not the best. But the odds of getting the most out of it will increase if one understands what biological aspects it tends to favour and why.

Die, selfish gene, die?

Throughout this book I have documented several instances where the gene's-eye view has stirred up strong emotions among biologists, philosophers, and laypeople. Such incidents are not just something of the past, but selfish genes are routinely used as the starting point by writers who want to pontificate about what is wrong with contemporary evolutionary theory. For example, in the fall of 2020, a special issue in *New Scientist* on 'the changing face of our greatest theory of nature' came with the splash 'Move over, selfish gene' (Barras 2020). In December 2013, the journalist David Dobbs applied a similar rhetorical approach in his long essay in *Aeon*, a digital magazine, titled 'Die, selfish gene, die' (Dobbs 2013). The gist of Dobbs's criticism was that new discoveries in gene regulation and developmental biology undermines the picture of evolution painted by the gene's-eye view. Dobbs succeeded in causing a stir and even got Dawkins to publish a reply on his foundation's website (Dawkins 2013b).

The criticism that evolutionary theory is failing to properly incorporate certain biological phenomena is not new (Welch 2017). Often, the complaint takes the form of lists of 'things in biology that interest me' (to paraphrase Clement Attlee's jibe against Winston Churchill's books on English history). Items on the lists may or may not have much to do with each other, and the same items may or may not appear on different lists. Other than developmental biology, things that often make an appearance on the list of what's wrong with evolutionary biology are phenotypic plasticity (Pigliucci 2001; West-Eberhard 2003), niche construction (Odling-Smee et al. 2003), and evolvability (Wagner 2005; ironically, Dawkins was one of the first to write about the 'evolution of evolvability' in Dawkins 1988).

There is an ongoing polemic whether these diverse observations demand an extended evolutionary synthesis (Pigliucci and Müller 2010; Laland et al. 2014, 2015; Charlesworth et al. 2017; Futuyma 2017; Huneman and Walsh 2017). What is at stake, and what according to its proponents needs an extension, is the modern synthesis of the 1930s and 1940s. The modern synthesis means different things to different people, and as a consequence critics have focused on different aspects, ranging from empirical to historical and philosophical ones (Gayon and Huneman 2019; Lewens 2019b). In an attempt to capture the essence of the modern synthesis, Huneman (2017) quotes Julian Huxley, a chief architect who coined the term in his book, the massive *Evolution: The Modern Synthesis* (Huxley 1942). In a letter to Ernst Mayr, Huxley described it as: 'Natural selection, acting on the heritable variation provided by the mutations and recombination of a Mendelian genetic constitution, is the main agency of biological evolution' (Huxley 1951, quoted in Huneman 2017, p. 71). The modern synthesis was more than the emergence of population genetics; it also included ecology, systematics, cytology, and palaeontology. Nevertheless, Huneman describes gene-centrism and adaptationism as the 'heart and soul' of the modern synthesis. This combination sounds very much like the gene's-eye view. That being said, biologists have long differed in whether they accept the whole package, and there are prominent examples of biologists accepting one but not the other. For example, Mayr endorsed natural selection as the main agent of evolution but rejected gene-centrism (Mayr 1959, 1983). At the opposite end, Michael Lynch strongly favours focusing on genes, writing that 'nothing in evolution makes sense except in light of population genetics', while also describing what he calls the 'Dawkins agenda to spread the word on the awesome power of natural selection' as 'misleading' (Lynch 2007b).

As noted above, both Williams and Dawkins considered their views to be capturing the insights of the modern synthesis. Regardless whether the modern synthesis at its core boils down to gene-centrism and adaptationism or not, recent and past debates suggest that there is more to the call for an extended evolutionary synthesis than just scientific disagreements over models and data. It also reflects a deeper division over how evolution should be construed. Take, for example, Sean B Carroll, the eloquent spokesperson for evolutionary developmental biology and critic of the gene's-eye view who wrote:

> Millions of biology students have been taught the view (from population genetics) that 'evolution is change in gene frequencies.' (…) This view forces the explanation toward mathematics and abstract descriptions of genes, and away from butterflies and zebras (…) The evolution of form is the main drama of life's story, both as found in the fossil record and in the diversity of living species. So, let's teach that story.
>
> (Carroll 2005, p. 294)

I personally find the story told by the gene's-eye view to be both dramatic and inspiring. It is my experience, however, that whether this picture is shared by others depend both on what biological questions that they are interested in, and how the gene's-eye view fits into the general intellectual environment where they received their scientific or philosophical training.

The gene's-eye view worldwide

Kim Sterelny once divided up evolutionary biology into what he called an American and a British tendency (Sterelny 2001, pp. 4–5). These two tendencies differ in their attitudes towards a number of issues. The British tendency has emphasized adaptation as the central problem of evolutionary biology, whereas the American has focused on diversity and its constraints. Following from this, the British tendency has given more weight to selective explanations, whereas the American has highlighted the importance of historical and chance explanations. This division is, of course, built on exaggeration and overgeneralization. For example, George Williams and Daniel Dennett are both Americans and would make great representatives for the British tendency. Nevertheless, in Chapter 1 I argued that the gene's-eye view grew out of a distinctly British intellectual environment. Anecdotally, its influence across the world appears to be heterogeneous.

Karl Popper aptly described science as 'knowledge without a knower' (Popper 1972, p. 109). The universe is out there and its properties does not change depending whether members of *Homo sapiens* can correctly describe them or not. From such a commitment to realism does not necessarily follow that there is only one way to describe those properties. Scientific theories are human constructs, which means one can ask questions about their origin and subsequent spread. For example, to what extent is the theory of evolution by natural selection a product of Darwin's quintessential British upper middle class upbringing? In a rather unusual book, two Darwin scholars, Robert Richards and Michael Ruse, debate this question (Richards and Ruse 2016). To Ruse, himself a Brit, the defining features that led to Darwin's theory were British in nature: the industrial revolution, notions of social progress, the Anglican Church, and thinkers like Robert Malthus, Adam Smith, and William Paley. In contrast, Richards paints the picture of a much more cosmopolitan Darwin: one deeply influenced by French, and especially German, contemporaries. Richards argues that rather than British Enlightenment figures, no other thinker had a stronger influence on the young Darwin than Alexander von Humboldt.

If Richards and Ruse deal with the origin of Darwin's theory, there is also a steady increase in the interest in so-called 'reception studies'. This research centres on how the theory of evolution was received in different countries, typically outside the European and American context. Examples in the genre include *China and Charles Darwin* (Pusey 1983), *From Man to Ape: Darwinism in Argentina, 1870–1920* (Novoa and Levine 2010), and *Darwin, Dharma, and the Divine: Evolutionary Theory and Religion in Modern Japan* (Godart 2017). To my knowledge, there have been no corresponding studies of the spread of the gene's-eye view.

One exception is Osamu Sakura's analysis of the reception of sociobiology in Japan (Sakura 1998). To many evolutionary biologists, Japan is primarily known for being the home of Motoo Kimura, who developed the neutral theory of molecular evolution (Kimura 1968). Much debated ever since its conception, neutral theory stimulated plenty of work and led to many theoretical and empirical advances (see Kern and Hahn 2018 and Jensen et al. 2019 for two contrasting takes on its current standing). As Sakura shows, however, Japanese evolutionary biology in the post World War II era, was influenced as least as much by the ecologist Kinji Imanishi, who was strongly critical of Darwin. Publishing almost exclusively in Japanese, and with great popular success, Imanishi rejected the idea that competition within and between species played any significant role in evolution and instead emphasized an inherent harmony in the living world (Halstead 1985). After the dominance of the Imanishi school, the arrival of sociobiology was a breath of fresh air. *The Selfish Gene* and *The Extended Phenotype* were both translated to Japanese, and so were *On Human Nature* (Wilson 1978) and *Evolution and the Theory of Games* (Maynard Smith 1982b). The younger generation of Japanese (evolutionary) ecologists rapidly incorporated these ideas.

I have discussed the reception and standing of the gene's-eye view with colleagues from various countries in Europe, South America, and Asia. These conversations suggest that the specific sub-field within evolutionary biology that has been dominant in a country matters for the standing of the gene's-eye view. Maynard Smith once remarked that he 'never knew a birdwatcher who was not a naive adaptationist' (Kohn 2004, p. 6). Countries where behavioural ecology has been strong also seemed to have embraced the gene's-eye view faster. This includes my native Sweden where bird-watching behavioural ecologists were instrumental not only in the professional organization of the field, such as the founding of the Evolutionary Biology Centre at Uppsala University, but also in the popularization of evolution (see, e.g., Ulfstrand 2008).

Evolutionary biology is a small field, which means that a few determined individuals may be able to leave an especially long-lasting legacy. Sakura (1998), for example, stresses the importance of the 'three musketeers' of Japanese sociobiology, Toshitaka Hidaka, Yoshiaki Itô, and Yûji Kishi for organizing translations of key books and for pushing the field forward. At this stage, these inferences from Sweden and Japan are all anecdotal but highlight the need for systematic culturally comparative reception studies.

Final thoughts

Throughout this book I have tried to guide the reader through the many-times quite intricate and sprawling literature that surrounds the gene's-eye view. Regardless of whether we consider it a Pareto error or the *primus inter pares* of all ways to think about evolution, I believe that understanding its origin, logic, and implications will help us realize its full potential.

I have often been frustrated by debates over the value of the gene's-eye view, both among scientific colleagues and in the public arena, and equally so by its supporters and its critics. It is my hope that this book can contribute some nuance to this discourse and that, after having made it this far, the reader will leave with a little bit familiarity with this special way to read nature.

References

Abbot P, Abe J, Alcock J, Alizon S, Alpedrinha JAC, Andersson M, et al. 2011. Inclusive fitness and eusociality. *Nature*. 471: E1–E4.

Ågren JA. 2013. Selfish genes and plant speciation. *Evolutionary Biology*. 40: 439–449.

Ågren JA. 2014. Evolutionary transitions in individuality: insights from transposable elements. *Trends in Ecology and Evolution*. 29: 90–96.

Ågren JA. 2016. Selfish genetic elements and the gene's-eye view of evolution. *Current Zoology*. 62: 659–665.

Ågren JA, Wang W, Koenig D, Neuffer B, Weigel D, and SI Wright. 2014. Mating system shifts and transposable element evolution in the plant genus *Capsella*. *BMC Genomics*. 15: 602.

Ågren JA and SI Wright. 2015. Selfish genetic elements and plant genome size evolution. *Trends in Plant Science*. 4: 195–196.

Ågren JA, Williamson RJ, Campitelli BE, and J Wheeler. 2017. Greenbeards in yeast: an undergraduate laboratory exercise to teach the genetics of cooperation. *Journal of Biological Education*. 51: 228–236.

Ågren JA and AG Clark. 2018. Selfish genetic elements. *PLoS Genetics*. 14: e1007700.

Ågren JA, Davies NG, and KR Foster. 2019a. Enforcement is central to the evolution of cooperation. *Nature Ecology and Evolution*. 3: 1018–1029.

Ågren JA, Munasinghe M, and AG Clark. 2019b. Sexual conflict through Mother's Curse and Father's Curse. *Theoretical Population Biology*. 129: 9–17.

Akçay E and J Van Cleve. 2016. There is no fitness but fitness, and the lineage is its bearer. *Philosophical Transactions of the Royal Society Series B*. 371: 20150085.

Aktipis A. 2020. *The Cheating Cell: How Evolution Helps Us Understand and Treat Cancer*. Princeton University Press.

Alexander RD and G Borgia. 1978. Group selection, altruism, and the levels of organization of life. *Annual Review of Ecology, Evolution, and Systematics*. 9: 449–474.

Allee WC, Emerson AE, Park O, Park T, and KP Schmidt. 1949. *Principles of Animal Ecology.* WB Saunders.

Allen B, Nowak MA, and EO Wilson. 2013. Limitations of inclusive fitness. *Proceedings of the National Academy of Sciences USA.* 110: 20135–20139.

Allen B and MA Nowak. 2016. There is no inclusive fitness at the level of the individual. *Current Opinion in Behavioral Sciences.* 12: 122–128.

Aminetzach YT, Macpherson JM, and DA Petrov. 2005. Pesticide resistance via transposition-mediated adaptive gene truncation in *Drosophila. Science.* 309: 764–767.

Anonymous. 1977. Review of *The Selfish Gene. The New York Times.* 17 March.

Ardrey R. 1970. *The Social Contract: A Personal Inquiry into the Evolutionary Sources of Order and Disorder.* Atheneum.

Aunger R, ed. 2001. *Darwinizing Culture: The Status of Memetics as a Science.* Oxford University Press.

Axelrod R and WD Hamilton. 1981. The evolution of cooperation. *Science.* 211: 1390–1396.

Ball P. 2018. Rethinking replication. *Chemistry Today.* 12 February.

Bar-Yam Y. 1999. Formalizing the gene-centered view of evolution. *Advances in Complex Systems.* 2: 277–281.

Barash DP. 1980. Evolutionary aspects of the family. In CK Hofling and JM Lewis, eds. *The Family: Evaluation and Treatment*, pp. 185–222. Brunnel–Mazel Publishers.

Barkow JH, Cosmides L, and J Tooby, eds. 1992. *The Adapted Mind: Evolutionary Psychology and the Generation of Culture.* Oxford University Press.

Barras C. 2020. Move over, selfish gene. *New Scientist.* 26 September.

Barry JM. 2005. *The Great Influenza: The Epic Story of the Deadliest Plague in History.* Penguin.

Bast J, Schaefer I, Schwander T, Maraun M, Scheu S, and K Kraaijeveld. 2016. No accumulation of transposable elements in asexual arthropods. *Molecular Biology and Evolution.* 33: 697–706.

Bast J, Jaron KS, Schuseil D, Roze D, and T Schwander. 2019. Asexual reproduction reduces transposable element load in experimental yeast populations. *eLife.* 8: e48548.

Bateson P. 1978. *The Selfish Gene* by Richard Dawkins. *Animal Behaviour.* 26: 316–318.

Bateson P. 1981. Sociobiology and genetic determinism. *Theoria to Theory.* 14: 291–300.

Bateson P. 1986. Sociobiology and human politics. In S Rose and L Appignanesi, eds. *Science and Beyond*, pp. 79–99. Basil Blackwell and The Institute of Contemporary Arts.

Bateson P. 1999. Ontogeny, communication and parent–offspring relationships. In S Bråten, ed. *Intersubjective Communication and Emotion in Early Ontogeny*, pp. 187–207. Cambridge University Press.

Bateson P. 2001. Behavioral development and Darwinian evolution. In S Oyama, P Griffiths, and R Gray, eds. *Cycles of Contingency*, pp. 149–166. MIT Press.

Bateson P. 2006a. The nest's tale: Affectionate disagreements with Richard Dawkins. In A Grafen and M Ridley, eds. *Richard Dawkins: How a Scientist Changed the Way We Think*, pp.164–175. Oxford University Press.

Bateson P. 2006b. The nest's tale: A reply to Richard Dawkins. *Biology and Philosophy*. 21: 553–558.

Bauer J. 2008. *Das kooperative Gen: Abschied vom Darwinismus*. Hoffman und Campe.

Belfort M and RP Bonocora. 2014. Homing endonucleases: From genetic anomalies to programmable genomic clippers. *Methods in Molecular Biology*. 1123: 1–26.

Bennett J. 1983. *Natural Selection, Heredity and Eugenics, including selected Correspondence of R.A. Fisher with Leonard Darwin and Others*. Clarendon Press.

Benetta ED, Akbari OS, and PM Ferree. 2019. Sequence expression of supernumerary B Chromosomes: Function or fluff? *Genes*. 10: 123.

Bentley MB. 2020. Group selection. In C Starr, ed. *Encyclopedia of Social Insects*. Springer.

Berdoy M, Webster JP, and DW Macdonald. 2000. Fatal attraction in rats infected with *Toxoplasma gondii*. *Proceedings of the Royal Society Series B*. 267: 1591–1594.

Beukeboom LW. 1994. Bewildering Bs: an impression of the 1st B-Chromosome Conference. *Heredity*. 73: 328–336.

Biernaskie JM, West SA, and A Gardner. 2011. Are greenbeards intragenomic outlaws? *Evolution*. 65: 2729–2742.

Biernaskie JM, Gardner A, and SA West. 2013. Multicoloured greenbeards, bacteriocin diversity and the rock-paper-scissors game. *Journal of Evolutionary Biology*. 26: 2081–2094.

Birch J. 2014. Hamilton's rule and its discontents. *British Journal for the Philosophy of Science*. 65: 381–411.

Birch J. 2017a. *The Philosophy of Social Evolution*. Oxford University Press.

Birch J. 2017b. The inclusive fitness controversy: finding a way forward. *Royal Society Open Science*. 4: 170335.

Birch J. 2019. Are kin and group selection rivals or friends? *Current Biology*. 29: R433–R438.

Birch J and S Okasha. 2015. Kin selection and its critics. *BioScience*. 65: 22–32.

Blackmore S. 1999. *The Meme Machine*. Oxford University Press.

Blumenstiel JP. 2011. Evolutionary dynamics of transposable elements in a small RNA world. *Trends in Genetics*. 27: 23–31.

Bonduriansky R and T Day. 2018. *Extended Heredity: A New Understanding of Inheritance of Evolution*. Princeton University Press.

Bonner JT. 1974. *On Development: The Biology of Form*. Harvard University Press.

Boomsma JJ. 2016. Fifty years of illumination about the natural levels of adaptation. *Current Biology*. 26: R1250–R1255.

Borello ME. 2005. The rise, fall and resurrection of group selection. *Endeavour*. 29: 43–47.

Borello ME. 2010. *Evolutionary Restraints: The Contentious History of Group Selection*. University of Chicago Press.

Bourke AFG. 2011. *Principles of Social Evolution*. Oxford University Press.

Bourke AFG. 2014. The gene's-eye view, major transitions and the formal darwinism project. *Biology and Philosophy*. 29: 241–248.

Bourrat P. 2021. *Conventions, Facts, and the Levels of Selection*. Cambridge University Press.

Boyd R and P Richerson. 1985. *Culture and the Evolutionary Process*. University of Chicago Press.

Brandon RN. 1990. *Adaptation and Environment*. Princeton University Press.

Brandon RN. 2019. Natural selection. In EN Zalta, ed. *The Stanford Encyclopedia of Philosophy*. https://plato.stanford.edu/archives/fall2019/entries/natural-selection/. Accessed on 19 June 2020.

Brandon RN and R Burian, eds. 1984. *Genes, Organisms, Populations: Controversies Over the Units of Selection*. MIT Press.

Brandon RN and HF Nijhout. 2006. The empirical nonequivalence of genic and genotypic models of selection: A (decisive) refutation of genic selectionism and pluralistic genic selectionism. *Philosophy of Science*. 73: 277–297.

Briggs A, Halvorson H, and A Steane. 2018. *It Keeps Me Seeking: The Invitation from Science, Philosophy, and Religion*. Oxford University Press.

Britten RJ and DE Kohne. 1968. Repeated sequences in DNA. *Science*. 161: 529–540.

Brosius J. 2005. Disparity, adaptation, exaptation, bookkeeping, and contingency at the genome level. *Paleobiology*. 31: 1–16.

Brookfield JFY. 2011. Host-parasite relationships in the genome. *BMC Biology*. 9: 67.

Brown A. 1999. *The Darwin Wars: The Scientific Battle for the Soul of Man*. Simon and Schuster.

Brown A. 2016. Genes and geniality: Dawkins, Midgley, and *The Selfish Gene*. In IJ Kidd and L McKinnell, eds. *Science and the Self: Animals, Evolution, and Ethics: Essays in Honour of Mary Midgley*, pp. 168–178. Routledge.

Bulmer MG. 2004. Did Jenkin's swamping argument invalidate Darwin's theory of natural selection? *British Journal for the History of Science*. 37: 281–297.

Burt A and R Trivers. 1998. Selfish DNA and breeding system in flowering plants. *Proceedings of the Royal Society Series B*. 265: 141–146.

Burt A and V Koufopanou. 2004. Homing endonuclease genes: the rise and fall and rise again of a selfish element. *Current Opinion in Genetics and Development*. 14: 609–615.

Burt A and R Trivers. 2006. *Genes in Conflict: The Biology of Selfish Genetic Elements*. Belknap Harvard University Press.

Buss LW. 1987. *The Evolution of Individuality*. Princeton University Press.

Cain A. 1964. The perfection of animals. In JD Carthy and CL Duddington, eds. *Viewpoints in Biology 3*, pp. 26–63. Butterworths.

Camus MF, Clancy DJ, and DK Dowling. 2012. Mitochondria, maternal inheritance, and male aging. *Current Biology*. 22: R1717–R1721.

Carroll SB. 2005. *Endless Forms Most Beautiful: The New Science of Evo Devo*. WW Norton.

Carroll SB. 2008. Evo-devo and an expanding evolutionary synthesis: A genetic theory of morphological evolution. *Cell*. 134: 25–36.

Case AL, Finseth FR, Barr CM, and L Fishman. 2016. Selfish evolution of cytonuclear hybrid incompatibility in Mimulus. *Proceedings of the Royal Society Series B*. 283: 20161493.

Cavalier-Smith T. 1980. How selfish is DNA? *Nature*. 285: 617–618.

Cavalli-Sforza LL and MW Feldman. 1978. Darwinian selection and 'altruism'. *Theoretical Population Biology*. 14: 268–280.

Cavalli-Sforza LL and M Feldman. 1981. *Cultural Transmission and Evolution: A Quantitative Approach*. Princeton University Press.

Champer J, Chung J, Lee YL, Liu C, Yang E, Wen Z, et al. 2019. Molecular safeguarding of CRISPR gene drive experiments. *eLife*. 8: e41439.

Charlesworth B. 1978. Some models of the evolution of altruistic behaviour between siblings. *Journal of Theoretical Biology*. 72: 297–319.

Charlesworth D, Barton NH, and B Charlesworth. 2017. The sources of adaptive variation. *Proceedings of the Royal Society Series B*. 284: 20162864.

Charnov EL. 1977. An elementary treatment of the genetical theory of kin selection. *Journal of Theoretical Biology*. 66: 541–550.

Claidière N and JB André. 2012. The transmission of genes and culture: A questionable analogy. *Evolutionary Biology*. 39: 12–24.

Clarke E. 2014. Origins of evolutionary transitions. *Journal of Biosciences*. 39: 303–317.

Clarke E. 2018. Adaptation, multilevel selection and organismality: A clash of perspectives. In R Joyce, ed. *The Routledge Handbook of Evolution and Philosophy*, pp. 35–48. Routledge.

Comfort N. 2015. Dawkins, redux. *Nature*. 525: 184–185.

Conway Morris S. 2008. Introduction. In S Conway Morris, ed. *The Deep Structure of Biology: Is Convergence Sufficiently Ubiquitous to Give a Directional Signal?*, pp. vii–x. Templeton Press.

Cordaux R and MA Batzer. 2006. Teaching an old dog new tricks: SINEs of canine genomic diversity. *Proceedings of the National Academy of Sciences USA*. 103: 1157–1158.

Cosmides LM and J Tooby. 1981. Cytoplasmic inheritance and intragenomic conflict. *Journal of Theoretical Biology*. 89: 83–129.

Crespi B and P Nosil P. 2013. Conflictual speciation: species formation via genomic conflict. *Trends in Ecology and Evolution*. 28: 48–57.

Cronin H. 1991. *The Ant and the Peacock: Altruism and Sexual Selection from Darwin to Today*. Cambridge University Press.

Cunningham C. 2010. *Darwin's Pious Idea: Why the Ultra-Darwinists and the Creationists Both Get it Wrong*. Eerdmans.

Curley JP, Davidson S, Bateson P, and FA Champagne. 2009. Social enrichment during postnatal development induces transgenerational effects on emotional and reproductive behavior in mice. *Frontiers in Behavioral Neuroscience*. 3: 25.

Daly M. 1982. Some caveats about cultural transmission models. *Human Ecology*. 10: 401–408.

Davies NB. 2000. *Cuckoos, Cowbirds and Other Cheats*. Poyser.

Davies NB, Krebs JR, and SA West. 2012. *An Introduction to Behavioural Ecology, 4th Edition*. Wiley-Blackwell.

Darwin C. 1838. Notebook M. *Darwin Online*. http://darwin-online.org.uk/content/frameset?viewtype=side&itemID=CUL-DAR125.-&pageseq=1. Accessed on 21 February 2020.

Darwin C. 1859. *On the Origin of Species by Means of Natural Selection, or the Preservation of Favoured Races in the Struggle for Life*. John Murray.

Darwin C. 1862. *On the Various Contrivances by Which British and Foreign Orchids Are Fertilised by Insects, and On the Good Effects of Intercrossing*. John Murray.

Darwin C. 1868. *The Variation of Animals and Plants under Domestication*. John Murray.

Darwin C. 1871. *The Descent of Man, and Selection in Relation to Sex*. John Murray.

Darwin C. 1887. In F Darwin, ed. *The Life and Letters of Charles Darwin, including an Autobiographical Chapter.* John Murray.

Darwin Correspondence Project. 2020a. Letter No. 2532: Letter from Charles Darwin to John Lubbock, 22 November 1859. https://www.darwinproject. ac.uk/letter/DCP-LETT-2532.xml. Accessed on 12 February 2020.

Darwin Correspondence Project 2020b. Letter No. 3259: Letter from Charles Darwin to John Murray 21 September 1861. https://www.darwinproject. ac.uk/letter/DCP-LETT-3259.xml. Accessed on 3 April 2020.

Darwin Correspondence Project. 2020c. Letter No. 729: Letter from Charles Darwin to Joseph Dalton Hooker, 11 January 1844. https://www. darwinproject.ac.uk/letter/DCP-LETT-729.xml. Accessed on 11 February 2020.

Darwin Correspondence Project. 2020d. Letter No. 1924: Letter from Charles Darwin to J.D. Hooker, 13 July 1856. https://www.darwinproject. ac.uk/letter/DCP-LETT-1924.xml. Accessed on 12 February 2020.

Darwin Correspondence Project 2020e. Letter No. 5140: Letter from A.R. Wallace to Charles Darwin, 2 July 1866. https://www.darwinproject. ac.uk/letter/DCP-LETT-5140.xml#back-mark-5140.f5. Accessed on 21 September 2020.

Dawkins R. 1966. *Selective Pecking in the Domestic Chick.* DPhil Dissertation, University of Oxford.

Dawkins R. 1976. *The Selfish Gene.* Oxford University Press.

Dawkins R. 1978. Replicator selection and the extended phenotype. *Zeitschrift für Tierpsychologie.* 47: 61–76.

Dawkins R. 1979. Twelve misunderstandings of kin selection. *Zeitschrift für Tierpsychologie.* 51: 184–200.

Dawkins R. 1981a. In defence of selfish genes. *Philosophy.* 56: 556–573.

Dawkins R. 1981b. Selfish genes in race or politics. *Nature.* 289: 528.

Dawkins R. 1982a. *The Extended Phenotype: The Gene as the Unit of Selection.* Oxford University Press.

Dawkins R. 1982b. Replicators and vehicles. In King's College Sociobiology Group, eds. *Current Problems in Sociobiology*, pp. 45–64. Cambridge University Press.

Dawkins R. 1985. Sociobiology: the debate continues. *New Scientist.* 24 January.

Dawkins R. 1986a. *The Blind Watchmaker.* Longman Scientific and Technical.

Dawkins R. 1986b. Sociobiology: The new storm in a tea cup. In S Rose and L Appignanesi, eds. *Science and Beyond*, pp. 61–78. Basil Blackwell and The Institute of Contemporary Arts.

Dawkins R. 1988. The evolution of evolvability. In CG Langton, ed. *Artificial Life: The Proceedings of an Interdisciplinary Workshop on the Synthesis and*

Simulation of Living Systems, pp. 201–220. Addison-Wesley Publishing Company.

Dawkins R. 1989. *The Selfish Gene 2ⁿᵈ Edition*. Oxford University Press.

Dawkins R. 1990. Parasites, desiderata lists and the paradox of the organism. *Parasitology*. 100: 63–67.

Dawkins R. 1993a. Foreword to the Canto Edition. In J Maynard Smith, *The Theory of Evolution*, pp. xi–xvi. Cambridge University Press.

Dawkins R. 1993b. Viruses of the mind. In B Dahlbom, ed. *Dennett and his Critics: Demystifying the Mind*, pp. 13–27. Blackwell.

Dawkins R. 1994a. The gene's-eye view of creation. *Nature 125th Anniversary Symposium: Our Place in Nature, the Evolution of our World*. https://www. youtube.com/watch?v=nfZMyJq6BBM&feature=youtu.be&t=1827. Accessed on 15 September 2020.

Dawkins R. 1994b. Burying the vehicle. *Behavioral and Brain Sciences*. 17: 616–617.

Dawkins R. 1995a. Reply to Lucy Sullivan. *Philosophical Transactions of the Royal Society Series B*. 349: 219–224.

Dawkins R. 1995b. *River Out of Eden: A Darwinian View of Life*. Basic Books.

Dawkins R. 1996. A survival machine. In J Brockman, ed. *The Third Culture*, pp. 75–86. Simon and Schuster.

Dawkins R. 1998a. Universal Darwinism. In D Hull and M Ruse, eds. *The Philosophy of Biology*, pp. 15–37. Oxford University Press.

Dawkins R. 1998b. *Unweaving the Rainbow: Science, Delusion and the Appetite for Wonder*. Houghton Mifflin.

Dawkins R. 2000. W.D. Hamilton obituary. *The Independent*. 10 March.

Dawkins R. 2003. *A Devil's Chaplain: Reflections on Hope, Lies, Science, and Love*. Houghton Mifflin.

Dawkins R. 2004a. *The Ancestor's Tale: A Pilgrimage to the Dawn of Life*. Weidenfeld and Nicolson.

Dawkins R. 2004b. Extended phenotype–but not too extended. A Reply to Laland, Turner and Jablonka. *Biology and Philosophy*. 19: 377–396.

Dawkins R. 2006a. Introduction to the 30ᵗʰ anniversary edition. In R Dawkins, *The Selfish Gene 30ᵗʰ Anniversary Edition*, pp. vii–xviii. Oxford University Press.

Dawkins R. 2006b [2007]. *The God Delusion* (paperback edition). Black Swan.

Dawkins R. 2008a. Evolution in biology tutoring? In D Palfreyman, ed. *The Oxford Tutorial: Thanks, You Taught Me How to Think*, pp. 50–54. Oxford Centre for Higher Education Policy Studies and Blackwell's.

Dawkins R. 2008b. The group delusion. *New Scientist*. 12 January.

Dawkins R. 2012a. The descent of Edward Wilson. *Prospect.* 24 May.

Dawkins R. 2012b. 'Group selection' is a cumbersome, time-wasting distraction. *Edge.* https://www.edge.org/conversation/the-false-allure-of-group-selection#rd. Accessed on 20 March 2020.

Dawkins R. 2012c. Foreword. In DP Hughes, J Brodeur, and F Thomas, eds. *Host Manipulation by Parasites*, pp. xi–xiii. Oxford University Press.

Dawkins R. 2013a. *An Appetite for Wonder: The Making of a Scientist.* Bantam Press.

Dawkins R. 2013b. Adversarial journalism and *The Selfish Gene. The Richard Dawkins Foundation for Reason and Science.* 6 December. https://www.richarddawkins.net/2013/12/adversarial-journalism-and-the-selfish-gene/target=_blank. Accessed on 13 July 2020.

Dawkins R. 2014. Inclusive fitness. *Presentation at the Oxford Union 18 February.* https://www.youtube.com/watch?v=MrgqUC7ZCxQ. Accessed on 14 July 2020.

Dawkins R. 2015a. *Brief Candle in the Dark: My Life in Science.* Random House.

Dawkins R. 2015b. Evolvability. In Brockman J, ed. *Life*, pp. 1–15. Harper Perennial.

Dawkins R. 2016. *The Extended Selfish Gene.* Oxford University Press.

Dawkins R. 2019. Foreword. In GC Williams, *Adaptation and Natural Selection: A Critique of Some Current Evolutionary Thought.* pp. ix–xiv. Princeton University Press.

Dawkins R and JR Krebs. 1978. Animal signals: information or manipulation? In JR Krebs and NB Davies, eds. *Behavioural Ecology: An Evolutionary Approach*, pp. 282–309. Blackwell Scientific.

De Tiège A, Van de Peer Y, Braeckman J, and KB Tanghe. 2018. The sociobiology of genes: the gene's eye view as a unifying behavioural-ecological framework for biological evolution. *History and Philosophy of the Life Sciences.* 40: 6.

De Tomaso AW. 2018. Allorecognition and stem cell parasitism: a tale of competition, selfish genes and greenbeards in a basal chordate. In P Pontarotti, ed. *Origin and Evolution of Biodiversity*, pp. 131–142. Springer.

Dennett DC. 1991. *Consciousness Explained.* Little, Brown and Co.

Dennett DC. 1993. 'Confusion over evolution': An exchange. *The New York Review of Books.* 14 January.

Dennett DC. 1995. *Darwin's Dangerous Idea: Evolution and the Meanings of Life.* Simon and Schuster.

Dennett DC. 1999. Afterword. In R Dawkins, *The Extended Phenotype: The Long Reach of the Gene* (paperback edition), pp. 265–268. Oxford University Press.

Dennett DC. 2011. Homunculi rule: Reflections on Darwinian Populations and natural selection by Peter Godfrey-Smith. *Biology and Philosophy*. 26: 475–488.

Deutsch D. 1997. *The Fabric of Reality: The Science of Parallel Universes and Its Implications.* Viking.

Dobbs D. 2013. Die, selfish gene, die. *Aeon*. 3 December. https://aeon.co/essays/the-selfish-gene-is-a-great-meme-too-bad-it-s-so-wrong. Accessed on 4 September 2020.

Dolgin ES and B Charlesworth. 2006. The fate of transposable elements in asexual populations. *Genetics*. 174: 817–827.

Doolittle WF. 1989. Hierarchical approaches to genome evolution. *Canadian Journal of Philosophy*. 102: 101–133.

Doolittle WF and C Sapienza. 1980. Selfish genes, the phenotype paradigm and genome evolution. *Nature*. 284: 601–603.

Douglas RN and JA Birchler. 2017. B chromosomes. In TA Bhat and A Ahmad, eds. *Chromosome Structure and Aberrations*, pp. 13–39. Springer.

Dover G. 1980. Ignorant DNA? *Nature*. 285: 618.

Dover G. 1999. Looping the evolutionary loop. *Nature*. 399: 217–218.

Dover G. 2000. Anti-Dawkins. In H Rose and S Rose, eds. *Alas Poor Darwin: Arguments against Evolutionary Psychology*, pp. 55–78. Harmony Books.

Dover G and WF Doolittle. 1980. Modes of genome evolution. *Nature*. 288: 646–647.

Eberhard WG. 1980. Evolutionary consequences of intracellular organelle competition. *Quarterly Review of Biology*. 55: 231–249.

Edwards AWF. 1994. The fundamental theorem of natural selection. *Biological Reviews*. 69: 443–474.

Edwards AWF. 1998. Natural selection and the sex ratio: Fisher's sources. *American Naturalist*. 151: 564–569.

Edwards AWF. 2014. R.A. Fisher's gene-centred view of evolution and the Fundamental Theorem of Natural Selection. *Biological Reviews*. 89: 135–147.

Eigen M. 1971. Self-organization of matter and the evolution of biological macromolecules. *Naturwissenschaften*. 58: 465–523.

El-Kirat-Chatel S, Beaussart A, Vincent SP, Abellán Flos M, Hols P, Lipke PN, and YF Dufrêne. 2015. Forces in yeast flocculation. *Nanoscale*. 7: 1760–1767.

Eldredge N. 1996. Comment on A Survival Machine by Richard Dawkins. In J Brockman, ed. *The Third Culture*, p. 91. Simon and Schuster.

Eldredge N. 2004. *Why We Do It: Rethinking Sex and The Selfish Gene.* WW Norton.

Eldredge N and SJ Gould. 1972. Punctuated equilibria: an alternative to phyletic gradualism. In TJM Schopf, ed. *Models in Paleobiology*, pp. 82–115. Freeman, Cooper and Co.

Elliott TA and TR Gregory. 2015. What's in a genome? The C-value enigma and the evolution of eukaryotic genome content. *Philosophical Transactions of the Royal Society Series B*. 370: 20140331.

Elmhirst S. 2015. Is Richard Dawkins destroying his reputation? *The Guardian*. 9 June.

Elsdon-Baker F. 2009. *The Selfish Genius: How Richard Dawkins Rewrote Darwin's Legacy*. Icon Books.

Esvelt KM, Smidler AL, Catteruccia F, and GM Church. 2014. Concerning RNA-guided gene drives for the alteration of wild populations. *eLife*. 3: e03401.

Ewens WJ. 1989. An interpretation and proof of the fundamental theorem of natural selection. *Theoretical Population Biology*. 36: 167–190.

Ewens WJ. 2011. What is the gene trying to do? *The British Journal for the Philosophy of Science*. 62: 155–176.

Falk R. 1986. What is a gene? *Studies in History and Philosophy of Science*. 17: 133–173.

Fanelli D. 2008. Altruism is no family matter. *New Scientist*. 9 January.

Faria GS, Varela SAM, and A Gardner. 2018. The relation between R.A. Fisher's sexy-son hypothesis and W.D. Hamilton's greenbeard effect. *Evolution Letters*. 2: 190–200.

Feldman MW and LL Cavalli-Sforza. 1981. Further remarks on Darwinian selection and 'altruism'. *Theoretical Population Biology*. 19: 251–260.

Feldman MW, Odling-Smee J, and KN Laland. 2017. Why Gupta et al.'s critique of niche construction theory is off target. *Journal of Genetics*. 96: 505–508.

Felsenstein J. 2012. Comment on 'The Paradox of Stasis'. *The Sandwalk Blog*. https://sandwalk.blogspot.com/2012/08/the-paradox-of-stasis.html?showComment=1344035878652#c2442008482190145215. Accessed on 21 February 2020.

Fisher RA. 1914. Some hopes of a eugenicist. *The Eugenics Review*. 5: 309–315.

Fisher RA. 1918. The correlation between relatives on the supposition of mendelian inheritance. *Transactions of the Royal Society of Edinburgh*. 52: 399–433.

Fisher RA. 1922. On the dominance ratio. *Proceedings of the Royal Society of Edinburgh*. 42: 321–341.

Fisher RA. 1927. On some objections to mimicry theory: statistical and genetic. *Transactions of the Royal Entomological Society of London*. 75: 269–278.

Fisher RA. 1930 [1999]. *The Genetical Theory of Natural Selection. A Complete Variorum Edition*. Oxford University Press.

Fisher RA. 1941. Average excess and average effect of a gene substitution. *Annals of Eugenics*. 11: 53–63.

Fisher RA. 1947. The renaissance of Darwinism. *The Listener*. 37: 1001 and 1009.

Fisher RA. 1954. Retrospect of criticisms of the theory of natural selection. In J Huxley, AC Hardy, and EB Ford, eds. *Evolution as a Process*, pp. 84–98. Allen and Unwin.

Fisher RA. 1958. Polymorphism and natural selection. *Journal of Ecology*. 46: 289–293.

Fisher RA. 1959. Natural selection from the genetical standpoint. *Australian Journal of Science*. 22 16–17.

Fisher RA and EB Ford. 1926. Variability of species. *Nature*. 118: 515–516.

Fisher Box J. 1978. *R.A. Fisher: The Life of a Scientist*. John Wiley and Sons.

Ford DH and RM Lerner. 1992. *Developmental Systems Theory: An Integrative Approach*. Sage.

Ford EB. 1964. *Ecological Genetics*. Chapman and Hall.

Francis RC. 2004. *Why Men Won't Ask for Directions: The Seductions of Sociobiology*. Princeton University Press.

Frank SA. 2003. Repression of competition and the evolution of cooperation. *Evolution*. 57: 693–705.

Frank SA. 2012. Natural selection. IV. The Price equation. *Journal of Evolutionary Biology*. 25: 1002–1019.

Frank SA and LD Hurst. 1996. Mitochondria and male disease. *Nature*. 383: 224.

Frank SA and BJ Crespi. 2011. Pathology from evolutionary conflict, with a theory of X chromosome versus autosome conflict over sexually antagonistic traits. *Proceedings of the National Academy of Sciences USA*. 108: 10886–10893.

Frank SA and MM Patten. 2020. Sexual antagonism leads to a mosaic of X–autosome conflict. *Evolution*. 74: 495–498.

Fredericksen MA, Zhang Y, Hazen ML, Loreto RG, Mangold CA, Chen DZ, and DP Hughes. 2017. Three-dimensional visualization and a deep-learning model reveal complex fungal parasite networks in behaviorally manipulated ants. *Proceedings of the National Academy of Sciences USA*. 114: 12590–12595.

Fromhage L and MD Jennions. 2019. The strategic reference gene: an organismal theory of inclusive fitness. *Proceedings of the Royal Society Series B*. 286: 20190459.

Futuyma DJ. 2017. Evolutionary biology today and the call for an extended synthesis. *Interface Focus.* 7: 20160145.

Futuyma DJ and SC Stearns. 2010. In Memoriam: George C. Williams. *Evolution.* 64: 3339–3343.

Gardner A. 2009. Adaptation as organism design. *Biology Letters.* 5: 861–864.

Gardner A. 2011. Kin selection under blending inheritance. *Journal of Theoretical Biology.* 284: 125–129.

Gardner A. 2014a. Genomic imprinting and the units of adaptation. *Heredity.* 113: 104–111.

Gardner A. 2014b. Life, the universe and everything. *Biology and Philosophy.* 29: 207–215.

Gardner A. 2015. Hamilton's rule. *American Naturalist.* 186: ii–iii.

Gardner A. 2016. The strategic revolution. *Cell.* 6: 1345–1348.

Gardner A. 2020. Price's equation made clear. *Philosophical Transactions of the Royal Society of London Series B.* 375: 20190361.

Gardner A, West SA, and NH Barton. 2007. The relation between multilocus population genetics and social evolution theory. *American Naturalist.* 169: 207–226.

Gardner A and A Grafen. 2009. Capturing the superorganism: a formal theory of group adaptation. *Journal of Evolutionary Biology.* 22: 659–671.

Gardner A and SA West. 2010. Greenbeards. *Evolution.* 64: 25–38.

Gardner A and JJ Welch. 2011. A formal theory of the selfish gene. *Journal of Evolutionary Biology.* 24: 1801–1813.

Gardner A, West SA, and G Wild. 2011. The genetical theory of kin selection. *Journal of Evolutionary Biology.* 24: 1020–1043.

Gardner A and F Úbeda. 2017. The meaning of intragenomic conflict. *Nature Ecology and Evolution.* 1: 1807–1815.

Gayon J and P Huneman. 2019. The modern synthesis: Theoretical or institutional event? *Journal of the History of Biology.* 52: 519–535.

Gemmell NJ, Metcalf VJ, and FW Allendorf. 2004. Mother's curse: the effect of mtDNA on individual fitness and population viability. *Trends in Ecology and Evolution.* 19: 238–244.

Gershenson S. 1928. A new sex-satio abnormality in *Drosophila obscura*. *Genetics.* 13: 488–507.

Ghiselin MT. 1997. *Metaphysics and the Origin of Species.* State University of New York Press.

Giberson K and M Artigas. 2007. *Oracles of Science: Celebrity Scientists versus God and Religion.* Oxford University Press.

Gillespie NC. 1990. Divine design and the industrial revolution: William Paley's abortive reform of natural theology. *Isis.* 81: 214–229.

Gintis H. 2014. Inclusive fitness and the sociobiology of the genome. *Biology and Philosophy*. 29: 477–515.

Gintis H. 2016. *Individuality and Entanglement: The Moral and Material Bases of Social Life*. Princeton University Press.

Godart GC. 2017. *Darwin, Dharma, and the Divine: Evolutionary Theory and Religion in Modern Japan*. University of Hawai'i Press.

Godfrey-Smith P. 2000a. Information, arbitrariness, and selection: Comments on Maynard Smith. *Philosophy of Science*. 67: 202–207.

Godfrey-Smith P. 2000b. The replicator in retrospect. *Biology and Philosophy*. 15: 403–423.

Godfrey-Smith P. 2009. *Darwinian Populations and Natural Selection*. Oxford University Press.

Godfrey-Smith P. 2011. Agents and acacias: Replies to Dennett, Sterelny, and Queller. *Biology and Philosophy*. 26: 501–515.

Goffman A. 2014. *On the Run: Fugitive Life in an American City*. University of Chicago Press.

Goodwin B. 1994 [2001]. *How the Leopard Changed Its Spots: The Evolution of Complexity*. Princeton University Press.

Gould SJ. 1977. Caring groups and selfish genes. *Natural History*. 86: 20–24.

Gould SJ. 1980. Is a new and general theory of evolution emerging? *Paleobiology*. 6: 1191–1130.

Gould SJ. 1983a. What happens to bodies if genes act for themselves? In SJ Gould, ed. *Hen's Teeth and Horse's Toes*, pp. 166–167. WW Norton.

Gould SJ. 1983b. The hardening of the modern synthesis. In M Grene, ed. *Dimensions of Darwinism; Themes and Counterthemes in Twentieth-Century Evolutionary Biology*, pp. 71–93. Cambridge University Press.

Gould SJ. 1992. The confusion over evolution. *The New York Review of Books*. 19 November.

Gould SJ. 1993. 'Confusion over evolution': An exchange. *The New York Review of Books*. 14 January.

Gould SJ. 1994. Tempo and mode in the macroevolutionary reconstruction of Darwinism. *Proceedings of the National Academy of Sciences USA*. 91: 6764–6771.

Gould SJ. 1997a. Darwinian fundamentalism. *The New York Review of Books*. June 12.

Gould SJ. 1997b. Evolution: The pleasures of pluralism. *The New York Review of Books*. June 26.

Gould SJ. 2002. *The Structure of Evolutionary Theory*. Belknap Harvard University Press.

Gould SJ and RC Lewontin. 1979. The spandrels of San Marco and the Panglossian paradigm: a critique of the adaptationist programme. *Proceedings of the Royal Society Series B*. 205: 581–598.

Gould SJ and EA Lloyd. 1999. Individuality and adaptation across levels of selection: How shall we name and generalize the unit of Darwinism? *Proceedings of the National Academy of Sciences USA*. 96: 11904–11909.

Grafen A. 1982. How not to measure inclusive fitness. *Nature*. 298: 425–426.

Grafen A. 1984. Natural selection, kin selection and group selection. In JR Krebs and NB Davies, eds. *Behavioural Ecology, 2nd Edition*, pp. 62–84. Blackwell Scientific.

Grafen A. 1998. Green beard as death warrant. *Nature*. 394: 521–522.

Grafen A. 1999. Formal Darwinism, the individual-as-maximizing-agent analogy, and bet-hedging. *Proceedings of the Royal Society Series B*. 266: 799–803.

Grafen A. 2002. A first formal link between the Price equation and an optimization program. *Journal of Theoretical Biology*. 217: 75–91.

Grafen A. 2003. Fisher the evolutionary biologist. *The Statistician*. 52: 319–329.

Grafen A. 2004. William Donald Hamilton. *Biographical Memoirs of Fellows of the Royal Society*. 50: 109–132.

Grafen A. 2007. The formal Darwinism project: A mid-term report. *Journal of Evolutionary Biology*. 20: 1243–1254.

Grafen A. 2008. The simplest formal argument for fitness optimization. *Journal of Genetics*. 87: 421–433.

Grafen A. 2014a. The formal Darwinism project in outline. *Biology and Philosophy*. 29: 155–174.

Grafen A. 2014b. The formal Darwinism project in outline: response to commentaries. *Biology and Philosophy*. 29: 281–292.

Grafen A. 2018. The left hand side of the fundamental theorem of natural selection. *Journal of Theoretical Biology*. 456: 175–189.

Grafen A and M Ridley, eds. 2006. *Richard Dawkins: How a Scientist Changed the Way We Think*. Oxford University Press.

Graff G. 1992. *Beyond the Culture Wars: How Teaching the Conflicts Can Revitalize American Education*. WW Norton.

Gray J. 2008. The atheist delusion. *The Guardian*. 15 March.

Gray R. 1992. Death of the gene: Developmental systems strike back. In P Griffiths, ed. *Trees of Life: Essays in Philosophy of Biology*, pp. 165–209. Kluwer Academic Publishers.

Gray R. 2001. Selfish genes or developmental systems? In R Singh, C Krimbas, D Paul, and J Beatty, eds. *Thinking About Evolution: Historical, Philosophical, and Political Perspectives*, pp. 184–207. Cambridge University Press.

Gregory TR. 2004. Macroevolution, hierarchy theory, and the C-value enigma. *Paleobiology*. 30: 179–202.

Gregory TR. 2013. Molecules and macroevolution: a Gouldian view of the genome. In GA Danieli, A Minelli, and T Pievani, eds. *Stephen J. Gould: The Scientific Legacy*, pp. 53–72. Springer.

Gregory TR, Elliott TA, and S Linquist. 2016. Why genomics needs multi-level evolutionary theory. In N Eldredge, T Pievani, E Serrelli, and I Temkin, eds. *Evolutionary Theory: A Hierarchical Perspective*, pp. 137–150. University of Chicago Press.

Greider G. 2014. *Den solidariska genen*. Ordfront.

Griesemer J. 2000. The units of evolutionary transition. *Selection*. 1: 67–80.

Griesemer J. 2006. Genetics from an evolutionary process perspective. In EM Neumann-Held and C Rehmann-Sutter, eds. *Genes in Development: Re-reading the Molecular Paradigm*, pp. 199–237. Duke University Press.

Griffiths PE and EM Neumann-Held. 1999. The many faces of the gene. *BioScience*. 49: 656–662.

Griffiths PE and RD Gray. 1994. Developmental systems and evolutionary explanation. *Journal of Philosophy*. 91: 277–304.

Griffiths PE and RD Gray. 1997. Replicator II – Judgement Day. *Biology and Philosophy*. 12: 471–492.

Griffiths P and K Sterelny. 1999. *Sex and Death: An Introduction to Philosophy of Biology*. University of Chicago Press.

Griffiths P and K Stotz. 2013. *Genetics and Philosophy: An Introduction*. Cambridge University Press.

Griffiths PE and J Taberny. 2013. Developmental systems theory: What does it explain, and how does it explain it? *Advances in Child Development and Behavior*. 44: 65–94.

Guilford T. 1985. Is kin selection involved in the evolution of warning coloration? *Oikos*. 45: 31–36.

Gupta M, Prasad NG, Dey S, Joshi A, and TN Vidya. 2017a. Niche construction in evolutionary theory: the construction of an academic niche? *Journal of Genetics*. 96: 491–504.

Gupta M, Prasad NG, Dey S, Joshi A, and TN Vidya. 2017b. Feldman et al. do protest too much, we think. *Journal of Genetics*. 96: 509–511.

Hacking I. 2012. Introductory essay. In TS Kuhn, *Structure of Scientific Revolutions: 50th Anniversary Edition*, pp. vii–xxxvii. University of Chicago Press.

Haig D. 1996. Gestational drive and the green-bearded placenta. *Proceedings of the National Academy of Sciences USA*. 93: 6547–6551.

Haig D. 1997. The social gene. In JR Krebs and NB Davies, eds. *Behavioural Ecology 4th Edition*, pp. 284–304. Blackwell Scientific.

Haig D. 2006a. The gene meme. In A Grafen and M Ridley, eds. *Richard Dawkins: How a Scientist Changed the Way We Think*, pp. 50–65. Oxford University Press.

Haig D. 2006b. Intragenomic politics. *Cytogenetic and Genome Research*. 113: 68–74.

Haig D. 2007. Weismann Rules! OK? Epigenetics and the Lamarckian temptation. *Biology and Philosophy*. 22: 415–428.

Haig D. 2012. The strategic gene. *Biology and Philosophy*. 27: 461–479.

Haig D. 2013. Imprinted green beards: a little less than kin and more than kind. *Biology Letters*. 9: 20130199.

Haig D. 2014. Genetic dissent and individual compromise. *Biology and Philosophy*. 29: 233–239.

Haig D. 2020. *From Darwin to Derrida: Selfish Genes, Social Selves, and the Meanings of Life*. MIT Press.

Haig D and A Grafen. 1991. Genetic scrambling as a defence against meiotic drive. *Journal of Theoretical Biology*. 153: 531–558.

Haldane JBS. 1932 [1990]. *The Causes of Evolution*. Princeton University Press.

Haldane JBS. 1955. Population genetics. *New Biology*. 18: 34–51.

Haldane JBS. 1964. A defense of beanbag genetics. *Perspectives in Biology and Medicine*. 7: 343–360.

Halstead B. 1985. Anti-darwinian theory in Japan. *Nature*. 317: 587–589.

Hamilton WD. 1963. The evolution of altruistic behavior. *American Naturalist*. 97: 354–356.

Hamilton WD. 1964a. The genetical evolution of social behaviour I. *Journal of Theoretical Biology*. 7: 1–16.

Hamilton WD. 1964b. The genetical evolution of social behaviour II. *Journal of Theoretical Biology*. 7: 17–52.

Hamilton WD. 1967. Extraordinary sex ratios. *Science*. 156: 477–488.

Hamilton WD. 1972. Altruism and related phenomena, mainly in social insects. *Annual Review of Ecology and Systematics*. 3: 193–232.

Hamilton WD. 1975. Innate social aptitudes of man: an approach from evolutionary genetics. In R Fox, ed. *Biosocial Anthropology*, pp. 133–155. Wiley.

Hamilton WD. 1977a. The play by nature. *Science*. 196: 757–759.

Hamilton WD. 1977b. *The Selfish Gene*. *Nature*. 267: 102.

Hamilton WD. 1996. *Narrow Roads of Gene Land, Volume 1: Evolution of Social Behaviour*. Oxford University Press.

Hamilton WD. 1999. Comment on *The Genetical Theory of Natural Selection. A Complete Variorum Edition*. Oxford University Press.

Hammerstein P. 1996. Darwinian adaptation, population genetics and the streetcar theory of evolution. *Journal of Mathematical Biology*. 34: 511–532.

Hampe M and SR Morgan. 1988. Two consequences of Richard Dawkins' view of genes and organisms. *Studies in History and Philosophy of Science Part A*. 19: 119–138.

Hancks DC and HH Kazazian Jr. 2016. Roles for retrotransposon insertions in human disease. *Mobile DNA*. 7: 9.

Hanke D. 2004. Teleology: the explanation that bedevils biology. In Cornwell J, ed. *Explanations: Styles of Explanation in Science*, pp. 143–155. Oxford University Press.

Hansell M. 2005. *Animal Architecture*. Oxford Univerity Press.

Harman O. 2011. *The Price of Altruism: George Price and the Search for the Origins of Kindness*. WW Norton.

Harrison E, MacLean RC, Koufopanou V, and A Burt. 2014. Sex drives intracellular conflict in yeast. *Journal of Evolutionary Biology*. 27: 1757–1763.

Hartfield M. 2015. Evolutionary genetic consequences of facultative sex and outcrossing. *Journal of Evolutionary Biology*. 29: 5–22.

Hartl DL and AG Clark. 2007. *Principles of Population Genetics 4th Edition*. Sinauer Associates.

Harvey PH and MD Pagel. 1991. *The Comparative Method in Evolutionary Biology*. Oxford University Press.

Havird JC, Forsythe ES, Williams AM, Werren JH, Dowling DK, and DB Sloan. 2019. Selfish mitonuclear conflict. *Current Biology*. 29: R496–R511.

Herrmann BG and H Bauer. 2012. The mouse t-haplotype: a selfish chromosome-genetics, molecular mechanism, and evolution. In M Macholán, SJE Baird, P Munclinger, and J Piálek, eds. *Evolution of the House Mouse*, pp. 297–314. Cambridge University Press.

Herron M. 2021. What are the major transitions? *Biology and Philosophy*. 36: 1–19.

Hickey DA. 1982. Selfish DNA: a sexually-transmitted nuclear parasite. *Genetics*. 101: 519–531.

Higgs PG and N Lehman. 2015. The RNA world: molecular cooperation at the origins of life. *Nature Reviews Genetics*. 16: 7–17.

Hill T, Schlötterer C, and AJ Betancourt. 2016. Hybrid dysgenesis in *Drosophila simulans* associated with a rapid invasion of the *P*-element. *PLoS Genetics*. 12: e1005920.

Hitchcock TJ and A Gardner. 2020. A gene's-eye view of sexual antagonism. *Proceedings of the Royal Society Series B*. 287: 20201633.

Hoover K, Grove M, Gardner M, Hughes DP, McNeil J, and J Slavicek. 2011. A gene for an extended phenotype. *Science*. 333: 1401.

Horgan J. 2015. Nobel laureate Steven Weinberg still dreams of final theory. *Cross Check Blog at Scientific American*. https://blogs.scientificamerican.com/cross-check/nobel-laureate-steven-weinberg-still-dreams-of-final-theory/. Accessed on 3 May 2020.

Hossenfelder S. 2018. *Lost in Math: How Beauty Leads Physis Astray*. Basic Books.

Hubby JL and RC Lewontin. 1966. A molecular approach to the study of genic heterozygosity in natural populations. I. The number of alleles at different loci in *Drosophila pseudoobscura*. *Genetics*. 54: 577–594.

Hughes DP. 2008. The extended phenotype within the colony and how it obscures social communication. In P d'Ettorre and DP Huges, ed. *Sociobiology of Communication: An Interdisciplinary Perspective*, pp. 171–190. Oxford University Press.

Hughes DP. 2012. Parasites and the superorganism. In DP Hughes, J Brodeur, and F Thomas, eds. *Host Manipulation by Parasites*, pp. 140–154. Oxford University Press.

Hughes D. 2013. Pathways to understanding the extended phenotype of parasites in their hosts. *Journal of Experimental Biology*. 216: 142–147.

Hughes DP, Andersen SB, Hywel-Jones NL, Himaman W, Billen J, and JJ Boomsma. 2011. Behavioral mechanisms and morphological symptoms of zombie ants dying from fungal infection. *BMC Ecology*. 11: 13.

Hughes DP, J Brodeur, and F Thomas, eds. 2012. *Host Manipulation by Parasites*. Oxford University Press.

Hull DL. 1980. Individuality and selection. *Annual Review of Ecology and Systematics*. 11: 311–332.

Hull DL. 1981. Units of evolution: a metaphysical essay. In UJ Jensen and R Harré, eds. *The Philosophy of Evolution*, pp. 23–44. The Harvester Press.

Hume D. 1779 [1990]. *Dialogues Concerning Natural Religion*. Penguin.

Hume D. 1793 [2004]. *A Treatise of Human Nature*. Penguin.

Huneman P and DM Walsh, eds. 2017. *Challenging the Modern Synthesis: Adaptation, Development, and Inheritance*. Oxford University Press.

Huneman P, ed. 2007. *Understanding Purpose: Kant and the Philosophy of Biology*. University Rochester Press.

Huneman P. 2017. Why would we call for a new evolutionary synthesis? The variation issue and the explanatory alternatives. In P Huneman and DM Walsh, eds. *Challenging the Modern Synthesis: Adaptation, Development, and Inheritance*, pp. 68–110. Oxford University Press.

Hurst GD and JH Werren. 2001. The role of selfish genetic elements in eukaryotic evolution. *Nature Reviews Genetics*. 2: 597–606.

Hurst LD. 1996. Adaptation and selection of genomic parasites. In MR Rose and GV Lauder, eds. *Adaptation*, pp. 407–449. Academic Press.

Huxley J. 1942 [2009]. *Evolution: The Modern Synthesis*. MIT Press.

Huxley TH. 1878 [1967]. Excerpt from The Genealogy of Animals. In C Bibby, ed. *The Essence of T.H. Huxley: Selections from his Writings*, pp. 169–170. St Martin's Press.

Huxley TH. 1888 [2001]. The struggle for existence in human society. In TH Huxley, *Evolution & Ethics and Other Essays, Collected Essays* Vol. 9, pp. 195–236. Adamant Media Corporation.

Huxley TH. 1894 [1989]. Evolution and ethics. In JG Paradis and GC Williams, eds. *Evolution and Ethics: T.H. Huxley's Evolution and Ethics with New Essays on its Victorian and Sociobiological Context*, pp. 57–174. Princeton University Press.

Ingold T. 2007. The trouble with 'evolutionary biology'. *Anthropology Today*. 23: 13–17.

Istvan MA Jr. 2013. Gould talking past Dawkins on the units of selection issue. *Studies in History and Philosophy of Biology and Biomedical Sciences*. 44: 327–335.

Jablonka E. 2004. From replicators to heritably varying phenotypic traits: The Extended Phenotype revisited. *Biology and Philosophy*. 19: 353–375.

Jablonka E and MJ Lamb. 2005. *Evolution in Four Dimensions: Genetics, Epigenetics, Behavioral, and Symbolic Variation in the History of Life*. MIT Press.

Jain HK. 1980. Incidental DNA. *Nature*. 288: 647–648.

James SM. 2010. *An Introduction to Evolutionary Ethics*. Wiley-Blackwell.

James W. 1880. Great men, great thoughts, and the environment. *Atlantic Monthly*. 66: 441–459.

Jamie GA. 2017. Signals, cues and the nature of mimicry. *Proceedings of the Royal Society Series B*. 284: 20162080.

Jansen VAA and M van Baalen. 2006. Altruism through beard chromo-dynamics. *Nature*. 440: 663–666.

Jensen JD, Payseur BA, Stephan W, Aquadro CF, Lynch M, Charlesworth D, and B Charlesworth. 2019. The importance of the neutral theory in 1968 and 50 years on: a response to Kern and Hahn 2018. *Evolution*. 73: 111–114.

Johannsen W. 1911. The genotype conception of heredity. *American Naturalist*. 45: 129–159.

Johnston C. 2014. Biological warfare flares up again between EO Wilson and Richard Dawkins. *The Guardian*. 7 November.

Kamath A. 2009. What is the unit of natural selection? Is the gene's eye view of evolution unacceptably reductionist? *Resonance*. 14: 1047–1059.

Kampourakis K. 2017. *Making Sense of Genes*. Cambridge University Press.

Kampourakis K. 2020. Why does it matter that many biology concepts are metaphors? In K Kampourakis and T Uller, eds. *Philosophy of Science for Biologists*, pp. 102–122. Cambridge University Press.

Kant I. 1790 [2008]. *Critique of Pure Judgement*. Oxford University Press.

Kapusta A, Suh A, and C Feschotte. 2017. Dynamics of genome size evolution in birds and mammals. *Proceedings of the National Academy of Sciences USA*. 8: 1460–1469.

Karlin S and C Matessi. 1983. Kin selection and altruism. *Proceedings of the Royal Society Series B.* 219: 327–353.

Kelleher ES, Barbash DA, and JP Blumenstiel. 2020. Taming the turmoil within: New insights on the containment of transposable elements. *Trends in Genetics.* 36: 474–489.

Keller L, ed. 1999. *Levels of Selection in Evolution.* Princeton University Press.

Keller L and KG Ross. 1998. Selfish genes: a green beard in the red fire ant. *Nature.* 394: 573–575.

Kern AD and MW Hahn. 2018. The neutral theory in light of natural selection. *Molecular Biology and Evolution.* 35: 1366–1371.

Kerr B and P Godfrey-Smith. 2002. Individualist and multi-level perspectives on selection in structured populations. *Biology and Philosophy.* 17: 477–517.

Ketcham R. 2018. Equivalence, interactors, and Lloyd's challenge to genic pluralism. In C Jeler, ed. *Multilevel Selection and the Theory of Evolution: Historical and Conceptual Issues,* pp. 71–98. Palgrave Pivot.

Kitcher P, Sterelny K, and CK Waters. 1990. The illusory riches of Sober's monism. *Journal of Philosophy.* 87: 158–161.

Kidwell MG. 1983. Evolution of hybrid dysgenesis determinants in *Drosophila melanogaster. Proceedings of the National Academy of Sciences USA.* 80: 1655–1659.

Kofler R, Hill T, Nolte V, Betancourt AJ, and C Schlötterer. 2015. The recent invasion of natural *Drosophila simulans* populations by the *P*-element. *Proceedings of the National Academy of Sciences USA.* 112: 6659–6663.

Kohn M. 2004. *A Reason for Everything: Natural Selection and the British Imagination.* Faber and Faber.

Korzybski A. 1933. *Science and Sanity: An Introduction to Non-Aristotelian Systems and General Semantics.* International Non-Aristotelian Library Publishing Company.

Krebs JR and NB Davies. 1993. *An Introduction to Behavioural Ecology 3rd Edition.* John Wiley and Sons.

Krebs JR and R Dawkins. 1984. Animal signals: mind-reading and manipulation. In JR Krebs and NB Davies, eds. *Behavioural Ecology: An Evolutionary Approach 2nd Edition,* pp. 380–402. Blackwell Scientific.

Kropotkin P. 1902 [2009]. *Mutual Aid: A Factor of Evolution.* Freedom Press.

Kruuk H. 2003. *Niko's Nature: A Life of Niko Tinbergen and his Science of Animal Behaviour.* Oxford University Press.

Lack D. 1954. *The Natural Regulation of Animal Numbers.* Clarendon Press.

Lack D. 1966. *Population Studies of Birds.* Clarendon Press.

Laland K. 2004. Extending the extended phenotype. *Biology and Philosophy.* 19: 313–325.

Laland K, Uller T, Feldman M, Sterelny K, Müller GB, Moczek A, et al. 2014. Does evolutionary theory need a rethink? *Nature*. 514: 161–164.

Laland KN, Uller T, Feldman MW, Sterelny K, Müller GB, Moczek A, Jablonka E, and J Odling-Smee. 2015. The extended evolutionary synthesis: its structure, assumptions and predictions. *Proceedings of the Royal Society Series B*. 282: 20151019.

Laland K, Matthews B, and MW Feldman. 2016. An introduction to niche construction theory. *Evolutionary Ecology*. 30: 191–202.

Lande R and SJ Arnold. 1983. The measurement selection on correlated characters. *Evolution*. 37: 1210–1226.

Langley CH. 1977. A little Darwinism. *BioScience*. 27: 692.

Larracuente AM and DC Presgraves. 2012. The selfish segregation distorter gene complex of *Drosophila melanogaster*. *Genetics*. 192: 33–53.

Laurent J. 1999. A note on the origin of memes/mnemes. *Journal of Memetics*. 3.

Lehmann L and F Rousset. 2020. When do individuals maximize their inclusive fitness? *American Naturalist*. 4: 717–732.

Leigh EG Jr. 1971. *Adaptation and Diversity: Natural History and the Mathematics of Evolution*. Freeman, Cooper and Co.

LeMahieu DL. 1976. *The Mind of William Paley: A Philosopher and His Age*. University of Nebraska Press.

Lessard S and WJ Ewens. 2019. The left-hand side of the Fundamental Theorem of Natural Selection: A reply. *Journal of Theoretical Biology*. 472: 77–83.

Lester R. 1981. Naming names. *Nature*. 293: 696.

Levin SR and A Grafen. 2019. Inclusive fitness is an indispensable approximation for understanding organismal design. *Evolution*. 73: 1066–1076.

Levin SR, Caro SM, Griffin AS, and SA West. 2019. Honest signaling and the double counting of inclusive fitness. *Evolution Letters*. 3: 428–433.

Levins R and RC Lewontin. 1985. *The Dialectical Biologist*. Harvard University Press.

Lewens T. 2004. *Organisms and Artifacts: Design in Nature and Elsewhere*. MIT Press.

Lewens T. 2015. *Cultural Evolution*. Oxford University Press.

Lewens T. 2018. Cultural evolution. In EN Zalta ed. *The Stanford Encyclopedia of Philosophy*. https://plato.stanford.edu/entries/evolution-cultural/. Accessed on 20 February 2020.

Lewens T. 2019a. Neo-Paleyan Biology. *Studies in History and Philosophy of Biological and Biomedical Sciences*. 76: 101185.

Lewens T. 2019b. The Extended Evolutionary Synthesis: what is the debate about, and what might success for the extenders look like? *Biological Journal of the Linnean Society*. 127: 707–721.

Lewis D. 1941. Male sterility in natural populations of hermaphrodite plants the equilibrium between females and hermaphrodites to be expected with different types of inheritance. *New Phytologist*. 46: 56–63.

Lewontin RC. 1961. Evolution and the theory of games. *Journal of Theoretical Biology*. 1: 382–403.

Lewontin RC. 1962. Interdeme selection controlling a polymorphism in the house mouse. *American Naturalist*. 96: 65–78.

Lewontin RC. 1966. Adaptation and natural selection. *Science*. 152: 338–339.

Lewontin RC. 1970. The units of selection. *Annual Review of Ecology and Systematics*. 1: 1–18.

Lewontin RC. 1972. Testing the theory of natural selection. *Nature*. 236: 181–182.

Lewontin RC. 1977a. *The Selfish Gene*. *Nature*. 266: 283–284.

Lewontin RC. 1977b. *The Selfish Gene*. *Nature*. 267: 202.

Lewontin RC. 1981. Credit due to Nabi. *Nature*. 291: 608.

Lewontin RC. 2001. In the beginning was the word. *Science*. 291: 1263–1264.

Lewontin RC and LC Dunn LC. 1960. The evolutionary dynamics of a polymorphism in the house mouse. *Genetics*. 45: 705–722.

Lewontin RC and JL Hubby. 1966. A molecular approach to the study of genic heterozygosity in natural populations. II. Amount of variation and degree of heterozygosity in *Drosophila pseudoobscura*. *Genetics*. 54: 595–609.

Lewontin RC, Rose S, and LJ Kamin. 1984. *Not in Our Genes: Biology, Ideology, and Human Nature*. Pantheon Books.

Liao X, Rong R, and DC Queller. 2015. Relatedness, conflict, and the evolution of eusociality. *PLoS Biology*. 13: e1002098.

Lindholm AK, Dyer KA, Firman RC, Fishman L, Forstmeier W, Holman L, et al. 2016. The ecology and evolutionary dynamics of meiotic drive. *Trends in Ecology and Evolution*. 31: 315–326.

Linquist S, Cottenie K, Elliott TA, Saylor B, Kremer SC, and TR Gregory. 2015. Applying ecological models to communities of genetic elements: the case of neutral theory. *Molecular Ecology*. 24: 3232–3242.

Lisch D. 2013. How important are transposons for plant evolution? *Nature Reviews Genetics*. 14: 49–61.

Lloyd EA. 1988 [1994]. *The Structure and Confirmation of Evolutionary Theory*. Princeton University Press.

Lloyd EA. 1989. A structural approach to defining units of selection. *Philosophy of Science*. 56: 395–418.

Lloyd EA. 1992. Unit of selection. In Keller EF and EA Lloyd, eds. *Keywords in Evolutionary Biology*, pp. 334–340. Harvard University Press.

Lloyd EA. 2005. Why the gene will not return. *Philosophy of Science*. 72: 287–310.

Lloyd EA. 2017. Units and levels of selection. In EN Zalta, ed. *The Stanford Encyclopedia of Philosophy*. https://plato.stanford.edu/entries/selection-units/. Accessed on 2 December 2019.

Lloyd EA, Dunn M, Cianciollo J, and C Mannouris. 2005. Pluralism without genic causes? *Philosophy of Science*. 72: 334–341.

Lorenz K. 1966. *On Aggression*. Methuen Publishing.

Lovelock JE. 1979. *Gaia: A New Look at Life on Earth*. Oxford University Press.

Lovelock JE and L Margulis. 1974. Atmospheric homeostasis by and for the biosphere: the Gaia hypothesis. *Tellus*. 24: 2–9.

Lu Q and P Bourrat. 2018. The evolutionary gene and the extended evolutionary synthesis. *British Journal for the Philosophy of Science*. 69: 775–800.

Lynch M. 2007a. *The Origins of Genome Architecture*. Sinauer Associate.

Lynch M. 2007b. The frailty of adaptive hypotheses for the origins of organismal complexity. *Proceedings of the National Academy of Sciences USA*. 104: 8597–8604.

Mackie JL. 1978. The law of the jungle. *Philosophy*. 53: 455–464.

Madgwick PG, Belcher LJ, and JB Wolf. 2019. Greenbeard genes: Theory and reality. *Trends in Ecology and Evolution*. 34: 1092–1103.

Markowetz F. 2017. All biology is computational biology. *PLoS Biology*. e2002050.

Marshall JAR. 2015. *Social Evolution and Inclusive Fitness Theory: An Introduction*. Princeton University Press.

Marshall R. 2016. From a biological point of view, and then some. *3amMagazine*. https://www.3-16am.co.uk/articles/from-a-biological-point-of-view-and-then-some?c=end-times-archive. Accessed on 4 January 2020.

Maynard Smith J. 1964. Group selection and kin selection. *Nature*. 201: 1145–1147.

Maynard Smith J. 1969. The status of neo-darwinism. In CH Waddington, ed. *Sketching Theoretical Biology*, pp. 82–89. Edinburgh University Press.

Maynard Smith J. 1975. Survival through suicide. *New Scientist*. 28 August.

Maynard Smith J. 1979. Thatcher's biology. *New Scientist*. 14 June.

Maynard Smith J. 1981. Genes and race. *Nature*. 289: 742.

Maynard Smith J. 1982a. Genes and memes. *London Review of Books*. 4 February.

Maynard Smith J. 1982b. *Evolution and the Theory of Games*. Cambridge University Press.

Maynard Smith J. 1985. The birth of sociobiology. *New Scientist*. 26 September.

Maynard Smith J. 1986. *The Problems of Biology*. Oxford University Press.

Maynard Smith J. 1987. How to model evolution. In J Dupré, ed. *The Latest on the Best: Essays on Evolution and Optimality*, pp. 119–131. MIT Press.

Maynard Smith J. 1993. 'Confusion over evolution': An exchange. *The New York Review of Books*. 14 January.

Maynard Smith J. 1995. Genes, memes, and minds. *The New York Review of Books*. November 30.

Maynard Smith J. 1997. Interview by Richard Dawkins. *Web of Stories*. https://www.webofstories.com/play/john.maynard.smith/40. Accessed on 2 June 2020.

Maynard Smith J. 1998. The units of selection. In GR Bock and JA Goode, eds. *The Limits of Reductionism in Biology*, pp. 203–210. Novartis Foundation/Wiley.

Maynard Smith J. 2000. The concept of information in biology. *Philosophy of Science*. 67: 177–194.

Maynard Smith J. 2002. Commentary on Kerr and Godfrey-Smith. *Biology and Philosophy*. 17: 523–527.

Maynard Smith J and G Price. 1973. The logic of animal conflict. *Nature*. 246: 15–18.

Maynard Smith J and E Szathmáry. 1995. *The Major Transitions in Evolution*. Oxford University Press.

Maynard Smith J and E Szathmáry. 1999. *The Origins of Life: From the Birth of Life to the Origin of Language*. Oxford University Press.

Mayr E. 1959. Where are we? *Cold Spring Harbor Symposium on Quantitative Biology*. 24: 1–14.

Mayr E. 1963. *Animal Species and Evolution*. Harvard University Press.

Mayr E. 1969. Footnotes on the philosophy of biology. *Philosophy of Science*. 36: 197–202.

Mayr E. 1975. The unity of the genotype. *Biologisches Zentralblatt*. 94: 377–588.

Mayr E. 1982. *The Growth of Biological Thought: Diversity, Evolution, and Inheritance*. Belknap Harvard University Press.

Mayr E. 1983. How to carry out the adaptationist program? *American Naturalist*. 121: 324–334.

Mayr E. 1991. *One Long Argument: Charles Darwin and the Genesis of Modern Evolutionary Thought*. Harvard University Press.

Mayr E. 1997. The objects of selection. *Proceedings of the National Academy of Sciences USA*. 94: 2091–2094.

Mayr E. 1998. *This Is Biology: The Science of the Living World*. Belknap Harvard University Press.

Mayr E. 2004. *What Makes Biology Unique? Considerations on the Autonomy of a Scientific Discipline*. Cambridge University Press.

McClintock B. 1950. The origin and behavior of mutable loci in maize. *Proceedings of the National Academy of Sciences USA*. 36: 344–355.

McClintock B. 1956. Controlling elements and the gene. *Cold Spring Harbor Symposium on Quantitative Biology*. 21: 197–216.

McGrath AE. 2011. *Darwinism and the Divine: Evolutionary Thought and Natural Theology*. Wiley-Blackwell.

McShea DW and Simpson C. 2011. The miscellaneous transitions in evolution. In B Calcott and K Sterelny, eds. *The Major Transitions Revisited*, pp. 99–106. MIT Press.

Medawar PB. 1981. Back to evolution. *The New York Review of Books*. 19 February.

Michod RE. 1999. *Darwinian Dynamics: Evolutionary Transitions in in Fitness and Individuality*. Princeton University Press.

Midgley M. 1978. *Beast and Man: The Roots of Human Nature*. Cornell University Press.

Midgley M. 1979. Gene juggling. *Philosophy*. 54: 439–458.

Midgley M. 1983. Selfish genes and social Darwinism. *Philosophy*. 58: 365–377.

Midgley M. 1985. *Evolution as Religion*. Routledge.

Midgley M. 2010. *The Solitary Self: Darwin and the Selfish Gene*. Acumen.

Milot E, Moreau C, Gagnon A, Cohen AA, Brais B, and D Labuda. 2017. Mother's curse neutralizes natural selection against a human genetic disease over three centuries. *Nature Ecology Evolution*. 1: 1400–1406.

Mitchell SD. 2003. *Biological Complexity and Integrative Pluralism*. Cambridge University Press.

Monod J. 1970. *Chance and Necessity: An Essay on the Natural Philosophy of Modern Biology*. Knopf.

Moore GE. 1903 [1993]. *Principia Ethica*. Cambridge University Press.

Moore J. 2002. *Parasites and the Behavior of Animals*. Oxford University Press.

Moran PAP and CAB Smith. 1966. Commentary on R.A. Fisher's paper on the Correlation between relatives on the supposition of Mendelian inheritance. *Eugenics Laboratory Memoirs, Volume XLI*. Cambridge University Press.

Morgan TH. 1935. The relation of genetics to physiology and medicine. *The Scientific Monthly*. 41: 5–18.

Moss L. 2003. *What Genes Can't Do*. MIT Press.

Murdoch I. 1953. *Sartre: Romantic Rationalist*. Yale University Press.

Müller GB. 2007. Evo-devo: Extending the evolutionary synthesis. *Nature Reviews Genetics*. 8: 943–949.

Nabi I. 1981a. Ethics of genetics. *Nature*. 290: 183.

Nabi I. 1981b. It wasn't me. *Nature*. 291: 374.

Nee S and J Maynard Smith. 1990. The evolutionary biology of molecular parasites. *Parasitology*. 100: 5–18.

Nesse RM. 2006. Why a lot of people with selfish genes are pretty nice – except for their hatred of *The Selfish Gene*. In Grafen A and M Ridley, eds. *Richard Dawkins: How a Scientist Changed the Way We Think*, pp. 203–212. Oxford University Press.

Nesse RM and GC Williams. 1994. *Why We Get Sick: The New Science of Darwinian Medicine*. Times Books.

Nicholson DJ and J Dupré, eds. 2018. *Everything Flows: Towards a Processual Philosophy of Biology*. Oxford University Press.

Niemann HJ. 2014. *Karl Popper and the Two New Secrets of Life: Including Karl Popper's Medawar Lecture 1986 and Three Related Texts*. Mohr Siebeck.

Nitecki MH and DV Nitecki, eds. 1993. *Evolutionary Ethics*. State University of New York Press.

Noble D. 2006. *The Music of Life: Biology Beyond Genes*. Oxford University Press.

Noble D. 2011. Neo-Darwinism, the Modern Synthesis and selfish genes: are they of use in physiology? *Journal of Physiology*. 589: 1007–1015.

Noble D. 2015. Evolution beyond neo-Darwinism: a new conceptual framework. *Journal of Experimental Biology*. 218: 7–13.

Noble D. 2018. Central dogma or central debate? *Physiology*. 33: 246–249.

Noble D and P Hunter. 2020. How to link genomics to physiology through epigenomics. *Epigenomics*. 12: 285–287.

Norton B and ES Pearson. 1976. A note on the background, to and refereeing of, R.A. Fisher's 1918 paper 'On the correlation between relatives on the supposition of mendelian inheritance'. *Notes and Records of the Royal Society of London*. 31: 151–162.

Novoa A and A Levine. 2010. *From Man to Ape: Darwinism in Argentina, 1870–1920*. University of Chicago Press.

Nowak MA. 2006. *Evolutionary Dynamics: Exploring the Equations of Life*. Belknap Harvard University Press.

Nowak M. 2015. Inclusive fitness. In J Brockman, ed. *This Idea Must Die: Scientific Theories That Are Blocking Progress*, pp. 443–446. Harper Perennial.

Nowak MA, Tarnita CE, and EO Wilson. 2010. The evolution of eusociality. *Nature*. 466: 1057–1062.

Nowak MA, CE Tarnita, and EO Wilson. 2011. A brief statement about inclusive fitness and eusociality. *Program for Evolutionary Dynamics*. http://ped.fas.harvard.edu/files/ped/files/inclusivefitness_statement_1_0.pdf. Accessed on 2 May 2020.

Nowak MA and B Allen. 2015. Inclusive fitness theorizing invokes phenomena that are not relevant for the evolution of eusociality. *PLoS Biology*. 13: e1002134.

Odling-Smee FJ. 1988. Niche-constructing phenotypes. In HC Plotkin, ed. *The Role of Behavior in Evolution*, pp. 73–132. MIT Press.

Odling-Smee FJ, Laland KN, and MW Feldman. 2003. *Niche Construction: The Neglected Process in Evolution*. Princeton University Press.

Okasha S. 2004. The 'averaging fallacy' and the levels of selection. *Biology and Philosophy*. 19: 167–184.

Okasha S. 2006. *Evolution and the Levels of Selection*. Oxford University Press.

Okasha S. 2008a. Fisher's fundamental theorem of natural selection—A philosophical analysis. *The British Journal for the Philosophy of Science*. 59: 319–351.

Okasha S. 2008b. The units and levels of selection. In S Sarkar and A Plutynski, eds. *A Companion to the Philosophy of Biology*, pp. 138–156. Wiley-Blackwell.

Okasha S. 2012. Social justice, genomic justice and the veil of ignorance: Harsanyi meets Mendel. *Economics and Philosophy*. 28: 43–71.

Okasha S. 2018. *Agents and Goals in Evolution*. Oxford University Press.

Okasha S. 2019. Reply to Dennett, Gardner and Rubin. *Metascience*. 28: 373–382.

Okasha S and C Paternotte. 2014. The formal Darwinism project: editors' introduction. *Biology and Philosophy*. 29: 153–154.

Olson ME, Arroyo-Santos A, and F Vergara-Silva. 2019. A user's guide to metaphors in ecology and evolution. *Trends in Ecology and Evolution*. 4: 605–615.

Orgel LE, Crick FHC, and C Sapienza. 1980. Selfish DNA. *Nature*. 288: 645–646.

Östergren G. 1945. Parasitic nature of extra fragment chromosomes. *Botaniska Notiser*. 2: 157–163.

Otsuka J. 2016. Causal foundations of evolutionary genetics. *The British Journal for the Philosophy of Science*. 67: 247–269.

Otsuka J. 2019. *The Role of Mathematic in Evolutionary Theory*. Cambridge University Press.

Otto SP and TA Day. 2007. *A Biologist's Guide to Mathematical Modeling in Ecology and Evolution*. Princeton University Press.

Oyama S. 1985. *The Ontogeny of Information: Developmental Systems and Evolution*. Cambridge University Press.

Paley W. 1785 [2013]. *Principles of Moral and Political Philosophy*. Cambridge University Press.

Paley W. 1794 [2009]. *Views of the Evidence of Christianity*. Cambridge University Press.

Paley W. 1802 [2008]. *Natural Theology Or, Evidences of the Existence and Attributes of the Deity, Collected from the Appearances of Nature*. Oxford University Press.

Paradis JG and GC Williams, eds. 1989. *Evolution and Ethics: T.H. Huxley's Evolution and Ethics with New Essays on its Victorian and Sociobiological Context*. Princeton University Press.

Patel MR, Miriyala GK, Littleton AJ, Yang H, Trinh K, et al. 2016. A mitochondrial DNA hypomorph of cytochrome oxidase specifically impairs male fertility in *Drosophila melanogaster*. *eLife*. 5 :e16923.

Paternotte C. 2020. Social evolution and the individual-as-maximising-agent analogy. *Studies in History and Philosophy of Biological and Biomedical Sciences*. 79: 101225.

Pathak DT, Wei X, Dey A, and D Wall. 2013. Molecular recognition by a polymorphic cell surface receptor governs cooperative behaviors in bacteria. *PLoS Genetics*. 9: e1003891.

Patten MM. 2018. Selfish X chromosomes and speciation. *Molecular Ecology*. 27: 3772–3782.

Patten MM. 2019. The X chromosome favors males under sexually antagonistic selection. *Evolution*. 73: 84–91.

Pauwels E. 2013. Mind the metaphor. *Nature*. 500: 523–524.

Peluffo AE. 2015. The 'Genetic Program': behind the genesis of an influential metaphor. *Genetics*. 200: 685–696.

Piel H. 2019. JMS in 15 images. *Helen Piel Blog*. https://helenpiel.wordpress.com/jms-in-15-images/#retirement. Accessed on 22 March 2020.

Pigliucci M. 2001. Beyond selfish genes. *Skeptical Inquirer*. 31: 20–21.

Pigliucci M. 2009. *Darwinian Populations and Natural Selection. Notre Dame Philosophical Reviews*. 15 August.

Pigliucci M. 2013. On the different ways of 'doing theory' in biology. *Biological Theory*. 7: 287–297.

Pigliucci M and G Müller, eds. 2010. *Evolution: the Extended Synthesis*. MIT Press.

Pinker S. 2012. The false allure of group selection. *Edge*. https://www.edge.org/conversation/steven_pinker-the-false-allure-of-group-selection. Accessed on 10 March 2020.

Pittendrigh CS. 1958. Adaptation, natural selection, and behavior. In A Roe and G Gaylord Simpson, eds. *Behavior and Evolution*, pp. 390–416. Yale University Press.

Popper KR. 1972. *Objective Knowledge: An Evolutionary Approach*. Oxford University Press.

Price GR. 1970. Selection and covariance. *Nature*. 227: 520–521.

Price GR. 1972. Fisher's 'fundamental theorem' made clear. *Annals of Human Genetics*. 36: 129–140.

Provine WB. 1971. *The Origins of Theoretical Population Genetics*. University of Chicago Press.

Provine WB. 1986. *Sewall Wright and Evolutionary Biology*. University of Chicago Press.

Pusey JR. 1983. *China and Charles Darwin*. Harvard University Press.

Queller DC. 1992. A general model for kin selection. *Evolution*. 46: 376–380.

Queller DC. 1996. The measurement and meaning of inclusive fitness. *Animal Behaviour*. 51: 229–232.

Queller DC. 1997. Cooperators since life began. *Quarterly Review of Biology*. 72: 184–188.

Queller DC. 2011. A gene's eye view of Darwinian populations. *Biology and Philosophy*. 26: 905–913.

Queller DC. 2017. Fundamental theorems of evolution. *American Naturalist*. 189: 345–353.

Queller D. 2019. What life is for: a commentary on Fromhage and Jennions. *Proceedings of the Royal Society Series B*. 286: 20191060.

Queller DC. 2020. The gene's eye view, the Gouldian knot, Fisherian swords and the causes of selection. *Philosophical Transactions of the Royal Society Series B*. 375: 20190354.

Queller DC, Ponte E, Bozzaro S, and JE Strassmann. 2003. Single-gene greenbeard effects in the social amoeba *Dictyostelium discoideum*. *Science*. 299: 105–106.

Rachels J. 1991. *Created from Animals: The Moral Implications of Darwinism*. Oxford University Press.

Reiss J. 2011. *Not by Design: Retiring Darwin's Watchmaker*. University of California Press.

Rhoades MM. 1942. Preferential segregation in maize. *Genetics*. 27: 395–407.

Rice WR. 2013. Nothing in genetics makes sense except in the light of genomic conflict. *Annual Review of Ecology, Evolution, and Systematics*. 44: 217–237.

Richards RJ and M Ruse. 2016. *Debating Darwin*. University of Chicago Press.

Ridley M and A Grafen. 1981. Are green beard genes outlaws? *Animal Behaviour*. 29: 954–955.

Ridley M. 2000. *Mendel's Demon: Gene Justice and the Complexity of Life*. Weidenfeld and Nicolson.

Ridley M. 2016. In retrospect: *The Selfish Gene*. *Nature*. 529: 462–463.

Rieseberg LH and BK Blackman. 2010. Speciation genes in plants. *Annals of Botany*. 106: 439–455.

Riley MA and JE Wertz. 2002. Bacteriocins: Evolution, ecology, and application. *Annual Review of Microbiology*. 56: 117–137.

Riskin J. 2016. *The Restless Clock: A History of the Centuries-Long Argument over What Makes Living Things Tick*. University of Chicago Press.

Robert JS, Hall BK, and WM Olson. 2001. Bridging the gap between developmental systems theory and evolutionary developmental biology. *BioEssays*. 23: 954–962.

Rodgers M. 2013. *Publishing and the Advancement of Science: From Selfish Genes to Galileo's Finger*. Imperial College Press.

Rodgers M. 2017. The story of *The Selfish Gene. Logos*. 28: 44–55.

Rodrik D. 2015. *Economics Rules: The Rights and Wrongs of the Dismal Science*. WW Norton and Company.

Roughgarden J. 2009. *The Genial Gene: Deconstructing Darwinian Selfishness*. University of California Press.

Rose S. 1981. Genes and race. *Nature*. 289: 335.

Rose S, ed. 1982a. *Against Biological Determinism*. Allison and Busby.

Rose S, ed. 1982b. *Towards a Liberatory Biology*. Allison and Busby.

Rosenberg A. 1983. Coefficients, effects, and genic selection. *Philosophy of Science*. 50: 332–338.

Rosenberg A. 1993. Genic selection, molecular biology and biological instrumentalism. *Midwest Studies in Philosophy*. 18: 343–362.

Rosenberg A. 2006. *Darwinian Reductionism: Or, How to Stop Worrying and Love Molecular Biology*. University of Chicago Press.

Rosenberg A. 2011. *An Atheist's Guide to Reality: Enjoying Life without Illusions*. WW Norton.

Ross KG and L Keller. 2002. Experimental conversion of colony social organization by manipulation of worker genotype composition in fire ants (*Solenopsis invicta*). *Behavioral Ecology and Sociobiology*. 51: 287–295.

Rothstein SI and DP Barash. 1983. Gene conflicts and the concepts of outlaw and sheriff alleles. *Journal of Social and Biological Structures*. 6: 367–379.

Ruse M. 1980. Charles Darwin and group selection. *Annals of Science*. 37: 615–630.

Ruse M. 1986. *Taking Darwin Seriously: A Naturalistic Approach to Philosophy*. Prometheus Books.

Ruse M. 1989. Teleology in biology: is it a cause for concern? *Trends in Ecology and Evolution*. 4: 51–54.

Ruse M. 2003. *Darwin and Design: Does Evolution have a Purpose?* Harvard University Press.

Ruse M. 2013. Science and the humanities: Stephen Jay Gould's quest to join the high table. *Science and Education.* 22: 2317–2326.

Ruse M. 2018. *On Purpose.* Princeton University Press.

Ruse M. 2019a. Removing god from biology. In P Harrison and JH Roberts, eds. *Science without God? Rethinking the History of Scientific Naturalism,* pp. 130–147. Oxford University Press.

Ruse M. 2019b. *A Meaning to Life.* Oxford University Press.

Sakura O. 1998. Similarities and varieties: A brief sketch on the reception of Darwinism and sociobiology in Japan. *Biology and Philosophy.* 13: 341–357.

Sahlins MD. 1976. *The Use and Abuse of Biology: An Anthropological Critique of Sociobiology.* University of Michigan Press.

Sanei M, Pickering R, Kumke K, Nasuda S, and A Houben. 2011. Loss of centromeric histone H3 (CENH3) from centromeres precedes uniparental chromosome elimination in interspecific barley hybrids. *Proceedings of the National Academy of Sciences USA.* 108: 498–505.

Sarkar S. 1994. The additivity of variance and the selection of alleles. *Proceedings of the Biennial Meeting of the Philosophy of Science Association.* 1: 3–12.

Sarkar S. 2000. Information in genetics and developmental biology: Comments on Maynard Smith. *Philosophy of Science.* 67: 208–213.

Schulze-Hagen K, Stokke BG, and TR Birkhead. 2008. Reproductive biology of the European Cuckoo *Cuculus canorus:* early insights, persistent errors and the acquisition of knowledge. *Journal of Ornithology.* 150: 1–16.

Scott TW. 2019. *Adaptation and Genetic Conflict.* DPhil Dissertation, University of Oxford.

Scott TW and SA West. 2019. Adaptation and the parliament of genes. *Nature Communications.* 10: 5163.

Scruton R. 2017. *On Human Nature.* Princeton University Press.

Seger J and P Harvey. 1980. The evolution of the genetical theory of social behaviour. *New Scientist.* 3 July.

Segerstråle U. 2000. *Defenders of the Truth: The Sociobiology Debate.* Oxford University Press.

Segerstrale U. 2013. *Nature's Oracle: The Life and Work of W.D. Hamilton.* Oxford University Press.

Serrato-Capuchina A and DR Matute. 2018. The role of transposable elements in speciation. *Genes.* 9: 254.

Servedio MR, Brandvain Y, Dhole S, Fitzpatrick CL, Goldberg EE, Stern CA, et al. 2014. Not just a theory—the utility of mathematical models in evolutionary biology. *PLoS Biology.* 12: e1002017.

Slatkin M. 1972. On treating the chromosome as the unit of selection. *Genetics.* 72: 157–168.

Slobodkin LB. 1966. The light and the way in evolution. *Quarterly Review of Biology*. 41: 191–194.

Smart JCC. 1959. Can biology be an exact science? *Synthese*. 11: 359–368.

Smolin L. 2006. *The Trouble with Physics: The Rise of String Theory, the Fall of a Science, and What Comes Next*. Houghton Mifflin Harcourt.

Smukalla S, Caldara M, Pochet N, Beauvais A, Guadagnini S, Yan C, et al. 2008. *FLO1* is a variable green beard gene that drives biofilm-like cooperation in budding yeast. *Cell*. 135: 726–537.

Sober E. 1984. *The Nature of Selection: Evolutionary Theory in Philosophical Focus*. MIT Press.

Sober E. 1990. The poverty of pluralism: A reply to Sterelny and Kitcher. *The Journal of Philosophy*. 87: 151–158.

Sober E. 2011. *Did Darwin Write the Origin Backwards? Philosophical Essays on Darwin's Theory*. Prometheus Books.

Sober E. 2020. AWF Edwards on phylogenetic inference, Fisher's theorem, and race. *Quarterly Review of Biology*. 95: 125–129.

Sober E and RC Lewontin. 1982. Artifact, cause and genic selection. *Philosophy of Science*. 49: 157–180.

Sober E and RC Lewontin. 1983. Reply to Rosenberg on genic selectionism. *Philosophy of Science*. 50: 648–650.

Sober E and DS Wilson. 1994. A critical review of philosophical work on the units of selection problem. *Philosophy of Science*. 61: 534–555.

Sober E and DS Wilson. 1998. *Unto Others: The Evolution and Psychology of Unselfish Behavior*. Harvard University Press.

Sober E and DS Wilson. 2011. *Adaptation and Natural Selection* revisited. *Journal of Evolutionary Biology*. 24: 462–468.

Soto AM and Sonnenschein C. 2020. Information, programme, signal: dead metaphors that negate the agency of organisms. *Interdisciplinary Science Reviews*. 45: 331–343.

Speers RA. 2012. A review of yeast flocculation. In RA Speers, ed. *Proceedings of the 2nd International Brewers Symposium: Yeast Flocculation, Vitality and Viability*, pp. 525–531. Master Brewers Associations of the Americas.

Sperber D. 2000. An objection to the memetic approach to culture. In Aunger R, ed. *Darwinizing Culture*, pp. 163–173. Oxford University Press.

Stamp Dawkins M. 2006. Living with *The Selfish Gene*. In A Grafen and M Ridley, eds. *Richard Dawkins: How a Scientist Changed the Way We Think*, pp. 45–49. Oxford University Press.

Stanley M. 2015. *Huxley's Church and Maxwell's Demon: From Theistic Science to Naturalistic Science*. University of Chicago Press.

Stearns SC. 2002. Less would have been more. *Evolution*. 56: 2339–2345.

Stearns SC. 2011. *George Christopher Williams 1926–2010, A Biographical Memoir.* National Academy of Sciences.

Stearns SC and R Medzhitov. 2015. *Evolutionary Medicine.* Oxford University Press.

Stent GS. 1977. You can take the ethics out of altruism but you can't take the altruism out of ethics. *Hastings Center Report.* 7: 33–36.

Sterelny K. 2000. The 'genetic program' program: A commentary on Maynard Smith on information in biology. *Philosophy of Science.* 67: 195–201.

Sterelny K. 2001. *The Evolution of Agency and Other Essays.* Cambridge University Press.

Sterelny K. 2006. Memes revisited. *British Journal for the Philosophy of Science.* 57: 145–165.

Sterelny K. 2007. *Dawkins vs. Gould: Survival of the Fittest.* Icon Books.

Sterelny K. 2011. Darwinian spaces: Peter Godfrey-Smith on selection and evolution. *Biology and Philosophy.* 26: 489–500.

Sterelny K and P Kitcher. 1988. The return of the gene. *Journal of Philosophy.* 85: 339–360.

Sterelny K, Smith KC, and M Dickison. 1996. The extended replicator. *Biology and Philosophy.* 11: 377–403.

Sterelny K and PE Griffiths. 1999. *Sex and Death: An Introduction to Philosophy of Biology.* University of Chicago Press.

Stewart-Williams S. 2018. *The Ape that Understood the Universe: How the Mind and Culture Evolve.* Cambridge University Press.

Stove D. 1992. A new religion. *Philosophy.* 67: 233–240.

Stove D. 1995. *Darwinian Fairytales: Selfish Genes, Errors of Heredity and Other Fables of Evolution.* Avebury.

Sturtevant AH. 1915. The behavior of the chromosomes as studied through linkage. *Zeitschrift für induktive Abstammungs- und Vererbungslehre.* 13: 234–287.

Sullivan LG. 1995. Myth, metaphor and hypothesis: how anthropomorphism defeats science. *Philosophical Transactions of the Royal Society Series B.* 349: 215–218.

Szathmáry E. 2006. The origin of replicators and reproducers. *Philosophical Transactions of the Royal Society Series B.* 361: 1761–1776.

Szitenberg A, Cha S, Opperman CH, Bird DM, Blaxter ML, and DH Lunt. 2016. Genetic drift, not life history or RNAi, determine long-term evolution of transposable elements. *Genome Biology and Evolution.* 8: 2964–2978.

Taylor PD and SA Frank. 1996. How to make a kin selection model. *Journal of Theoretical Biology.* 180: 27–37.

Tenaillon MI, Hollister JD, and BS Gaut. 2010. A triptych of the evolution of plant transposable elements. *Trends in Plant Science*. 15: 471–478.

Tooby J and LM Cosmides. 1997. Unpublished letter to the Editor of *The New York Review of Books* on Stephen Jay Gould's 'Darwinian Fundamentalism' (June 12, 1997) and 'Evolution: The Pleasures of Pluralism' (June 26, 1997). *CogWeb*. http://cogweb.ucla.edu/Debate/CEP_Gould.html#1. Accessed on 9 August 2020.

Trivers RL. 1971. The evolution of reciprocal altruism. *Quarterly Review of Biology*. 46: 35–57.

Trivers RL. 1974. Parent-offspring conflict. *American Zoologist*. 14: 249–264.

Trivers R. 2009. Genetic conflict within the individual. *Sonderdruck der Berliner-Brandenburgische Akademie der Wissenschschaften*. 14: 149–199.

Trivers R. 2015. *Wild Life: Adventures of an Evolutionary Biologist*. Biosocial Research.

Trivers R, Burt A, and BG Palestis. 2004. B chromosomes and genome size in flowering plants. *Genome*. 47: 1–8.

Turner JS. 2004. Extended phenotypes and extended organisms. *Biology and Philosophy*. 19: 327–352.

Ulfstrand S. 2008. *Darwins idé: den bästa idé någon någonsin haft och hur den fungerar idag*. Symposium.

Uller T and KN Laland, eds. 2019. *Evolutionary Causation: Biological and Philosophical Reflections*. MIT Press.

van Valen L. 1981. Nabi—A life. *Nature*. 293: 422.

van Veelen M, Allen B, Hoffman M, Simon B, and C Veller. 2016. Hamilton's rule. *Journal of Theoretical Biology*. 414: 176–230.

van't Hof AE, Campagne P, Rigden DJ, Yung CJ, Lingley J, Quail MA, Hall N, Darby AC, and IJ Saccheri. 2016. The industrial melanism mutation in British peppered moths is a transposable element. *Nature*. 534: 102–105.

Venner S, Feschotte C, and C Biemont. 2009. Dynamics of transposable elements: towards a community ecology of the genome. *Trends in Genetics*. 25: 317–323.

Verspoor RL, Smith JML, Mannion NML, Hurst GDD, and TAR Price. 2018. Strong hybrid male incompatibilities impede the spread of a selfish chromosome between populations of a fly. *Evolution Letters*. 2: 169–179.

von Frisch K. 1974. *Animal Architecture*. Harcourt.

Vrba ES and N Eldredge. 1984. Individuals, hierarchies and processes: towards a more complete evolutionary theory. *Paleobiology*. 10: 146–171.

Waddington CH. 1975. Mindless societies. *The New York Review of Books*. 7 August.

Wade MJ. 1978. *The Selfish Gene*. *Evolution*. 32: 220–221.

Wade MJ. 1992. Sewall Wright: gene interaction and the shifting balance theory. In D Futuyma and J Antonovics, eds. *Oxford Surveys in Evolutionary Biology Volume 8*, pp. 35–64. Oxford University Press.

Wade MJ. 2002. A gene's eye view of epistasis, selection and speciation. *Journal of Evolutionary Biology*. 15: 337–346.

Wade MJ. 2016. *Adaptation in Metapopulations: How Interaction Changes Evolution*. University of Chicago Press.

Wade MJ and DM Drown. 2016. Nuclear–mitochondrial epistasis: a gene's eye view of genomic conflict. *Ecology and Evolution*. 6: 6460–6472.

Wagner A. 2005. *Robustness and Evolvability in Living Systems*. Princeton University Press.

Wagner A. 2009. Transposable elements as genomic diseases. *Molecular Biosystems*. 5: 32.

Walsh B and M Lynch. 2018. *Evolution and Selection of Quantitative Traits*. Oxford University Press.

Walsh DM. 2004. Bookkeeping or metaphysics? The units of selection debate. *Synthese*. 138: 337–361.

Walsh DM. 2015. *Organisms, Agency, and Evolution*. Cambridge University Press.

Washburn SL. 1978. Human behavior and the behavior of other animals. *American Psychologist*. 33: 405–418.

Waters CK. 1986. *Models of Natural Selection from Darwin to Dawkins*. PhD Dissertation, Indiana University.

Waters CK. 1991. Tempered realism about the forces of selection. *Philosophy of Science*. 58: 553–573.

Waters CK. 2005. Why genic and multilevel selection theories are here to stay. *Philosophy of Science*. 72: 311–333.

Weber JN, Peterson BK, and HE Hoekstra. 2013. Discrete genetic modules are responsible for complex burrow evolution in *Peromyscus* mice. *Nature*. 493: 402–405.

Weinberger N. 2011. Is there an empirical disagreement between genic and genotypic selection models? A response to Brandon and Nijhout. *Philosophy of Science*. 78: 225–237.

Weismann A. 1892. *Das Keimplasma: eine Theorie der Vererbung*. Verlag von Gustav Fischer.

Welch JJ. 2017. What's wrong with evolutionary biology? *Biology and Philosophy*. 32: 263–279.

Werren JH. 2011. Selfish genetic elements, genetic conflict, and evolutionary innovation. *Proceedings of the National Academy of Sciences USA*. 108: 10863–10870.

Werren JH, Nur U, and CI Wu. 1988. Selfish genetic elements. *Trends in Ecology and Evolution.* 3: 297–302.

West SA, Griffin AS, and A Gardner. 2007. Evolutionary explanations for cooperation. *Current Biology.* 17: R661–R672.

West SA and A Gardner. 2010. Altruism, spite, and greenbeards. *Science.* 327: 1341–1344.

West SA and A Gardner. 2013. Adaptation and inclusive fitness. *Current Biology.* 22: R577–R584.

West SA, Fisher RM, Gardner A, and ET Kiers. 2015. Major evolutionary transitions in individuality. *Proceedings of the National Academy of Sciences USA.* 112: 10112–10119.

West-Eberhard MJ. 2003. *Developmental Plasticity and Evolution.* Oxford University Press.

Williams B. 1976. Review: *The Selfish Gene. New Scientist.* 4 November.

Williams GC. 1957. Pleiotropy, natural selection, and the evolution of senescence. *Evolution.* 11: 398–411.

Williams GC. 1966. *Adaptation and Natural Selection: A Critique of Some Current Evolutionary Thought.* Princeton University Press.

Williams GC. 1985. A defense of reductionism in evolutionary biology. In R Dawkins and M Ridley, eds. *Oxford Surveys in Evolutionary Biology Volume 2,* pp. 1–27. Oxford University Press.

Williams GC. 1989. A sociobiological expansion of evolution and ethics. In JG Paradis and GC Williams, eds. *Evolution and Ethics: T.H. Huxley's Evolution and Ethics with New Essays on its Victorian and Sociobiological Context,* pp. 179–214. Princeton University Press.

Williams GC. 1992. *Natural Selection: Domains, Levels, and Challenges.* Oxford University Press.

Williams GC. 1993. Mother nature is a wicked old witch. In MH Nitecki and CV Nitecki, eds. *Evolutionary Ethics,* pp. 217–231. State University of New York Press.

Williams GC. 1996a. *Adaptation and Natural Selection: A Critique of Some Current Evolutionary Thought, 30th Anniversary Edition.* Princeton University Press.

Williams GC. 1996b. A package of information. In J Brockman, ed. *The Third Culture: Beyond the Scientific Revolution,* pp. 38–50. Simon and Schuster.

Williams GC. 1997. *The Pony Fish's Glow: And other Clues to Plan and Purpose in Nature.* Basic Books.

Williams GC and DC Williams. 1957. Natural selection of individually harmful social adaptations among sibs with special reference to social insects. *Evolution.* 11: 32–39.

Wilson DS. 2015a. *Does Altruism Exist? Culture, Genes, and the Welfare of Others.* Yale University Press.

Wilson DS. 2015b. The spandrels of San Marco Revisited: An interview with Richard C. Lewontin. *This View of Life.* https://evolution-institute.org/the-spandrels-of-san-marco-revisited-an-interview-with-richard-c-lewontin/. Accessed on 19 January 2020.

Wilson DS and EO Wilson. 2007. Rethinking the theoretical foundation of sociobiology. *Quarterly Review of Biology.* 82: 327–348.

Wilson DS and EO Wilson. 2008. Evolution for the good of the group. *American Scientist.* 96: 380–389.

Wilson EB. 1907. The supernumerary chromosomes of *Hemiptera. Science.* 26: 870–871.

Wilson EO. 1975. *Sociobiology: The New Synthesis.* Harvard University Press.

Wilson EO. 1978. *On Human Nature.* Harvard University Press.

Wilson EO. 1981. Who is Nabi? *Nature.* 290: 623.

Wilson EO. 2008. One giant leap: How insects achieved altruism and colonial life. *BioScience.* 58: 17–25.

Wilson EO. 2012. *The Social Conquest of Earth.* WW Norton.

Wilson RA. 2005. *Genes and the Agents of Life: The Individual in the Fragile Sciences: Biology.* Cambridge University Press.

Wimsatt WC. 1970. *Adaptation and Natural selection: A Critique of Some Current Evolutionary Thought.* George C. Williams. *Philosophy of Science.* 37: 620–623.

Wimsatt WC. 1980. Reductionistic research strategies and their biases in the units of selection controversy. In T Nickles, ed. *Scientific Discovery, Vol. 2, CaseStudies.* Reidel.

Wimsatt WC. 1999. Genes, memes, and cultural heredity. *Biology and Philosophy.* 14: 279–310.

Winnie JA, 2000. Information and structure in molecular biology: Comments on Maynard Smith. *Philosophy of Science.* 67: 517–526.

Winther RG, Wade MJ, and CC Dimond. 2013. Pluralism in evolutionary controversies: styles and averaging strategies in hierarchical selection theories. *Biology and Philosophy.* 28: 957–979.

Woit P. 2006. *Not Even Wrong: The Failure of String Theory and the Search for Unity in Physical Law.* Basic Books.

Wright S. 1930. *The Genetical Theory of Natural Selection*: A review. *Journal of Heredity.* 21: 349–356.

Wright S. 1931. Evolution in Mendelian populations. *Genetics.* 16: 97–159.

Wright S. 1980. Genic and organismic selection. *Evolution.* 34: 825–843.

Wright SI and DJ Schoen. 1999. Transposon dynamics and the breeding system. *Genetica.* 107: 139–148.

Wynne-Edwards VC. 1962. *Animal Dispersion in Relation to Social Behaviour.* Oliver and Boyd.

Wynne-Edwards VC. 1963. Intergroup selection in the evolution of social systems. *Nature.* 200: 623–626.

Yanai I and M Lercher. 2016. *The Society of Genes.* Harvard University Press.

Yanai I and M Lercher. 2020. The two languages of science. *Genome Biology.* 21: 147.

Zuk M. WD Hamilton. *Science.* 218: 384–387.

Zurita S, Cabrero J, López-León MD, and JPM Camacho. 1998. Polymorphism regeneration for a neutralized selfish B chromosome. *Evolution.* 52: 274–277.

Index

Tables and boxes are indicated by an italic *t* and *b* following the page number.